SICK BUILDING SYNDROME

SICK BUILDING SYNDROME

SYNDROME

Sources, Health Effects, Mitigation

by

M.C. Baechler, D.L. Hadley, T.J. Marseille, R.D. Stenner

Pacific Northwest Laboratory
Portland, Oregon

M.R. Peterson, D.F. Naugle

Research Triangle Institute
Research Triangle Park, North Carolina

M.A. Berry

U.S. Environmental Protection Agency
Research Triangle Park, North Carolina

NOYES DATA CORPORATION

Park Ridge, New Jersey, U.S.A.

Published in the United States of America by
Noyes Data Corporation
Mill Road, Park Ridge, New Jersey 07656

10 9 8 7 6 5 4 3 2 1

Library of Congress Cataloging-in-Publication Data

Sick building syndrome : sources, health effects, mitigation / by M.C.
        Baechler . . . [et al.].
                p.   cm. -- (Pollution technology review, ISSN 0090-516X ; no.
        205)
            Includes bibliographical references and index.
            ISBN 0-8155-1289-9 :
            1. Indoor air pollution--Health aspects. 2. Organic compounds-
        -Toxicology. 3. Buildings--Health aspects. I. Baechler, M.C.
        II. Series.
        RA577.5.S53   1991
        615.9'02--dc20   91-20423
                                                      CIP

Foreword

This book discusses the aspect of indoor air pollution referred to as "Sick Building Syndrome." Covered are sources and health effects of various indoor air pollutants, and methods for mitigation of the problem, plus suggested analytical methods for environmental carcinogens found in indoor air.

The quality of indoor air in commercial buildings is dependent on the complex interaction between sources of indoor pollutants, environmental factors within buildings such as temperature and humidity, the removal of air pollutants by air-cleaning devices, and the removal and dilution of pollutants from outside air.

Indoor air quality can be impacted by hundreds of different chemicals. More than 900 different organic compounds alone have been identified in indoor air. Health effects that could arise from exposure to individual pollutants or mixtures of pollutants cover the full range of acute and chronic effects, including largely reversible responses, such as rashes and irritations, to the irreversible toxic and carcinogenic effects. These indoor contaminants are emitted from a large variety of materials and substances that are widespread components of everyday life.

Information concerning indoor air quality and mitigation of the problem in both new and existing commercial buildings is summarized in Part I of the book. Sick building syndrome and specific pollutants are discussed, including mitigation techniques, ventilation, and the interaction between energy conservation activities and indoor air quality.

Part II reflects information currently available on the health effects associated with each of the selected contaminants. This information was obtained by conducting a comprehensive literature search on each of the selected contaminants. In addition, three active, professionally maintained and peer-reviewed on-line chemical databases were used to locate the latest health effects information on each contaminant. Contaminants discussed are those that have been measured in the indoor air of a public building, or have been measured (significant concentration) in test situations simulating indoor air quality (as presented in the referenced literature), and have a significant hazard rating.

Part III presents suggested sampling procedures and analytical approaches for known and suspected carcinogens found in indoor air.

The information in the book is from the following documents:

> *Indoor Air Quality Issues Related to the Acquisition of Conservation in Commercial Buildings,* prepared by M.C. Baechler, D.L. Hadley, and T.J. Marseille of Pacific Northwest Laboratory (operated by Battelle Memorial Institute for the U.S. Department of Energy) for Bonneville Power Administration under an agreement with the U.S. Department of Energy, September 1990.

> *Health Effects Associated with Energy Conservation Measures in Commercial Buildings–Volume 2: Review of the Literature,* prepared by R.D. Stenner and M.C. Baechler of Pacific Northwest Laboratory (operated by Battelle Memorial Institute for the U.S. Department of Energy) for Bonneville Power Administration under an agreement with the U.S. Department of Energy, September 1990.

> *Indoor Air–Assessment: Methods of Analysis for Environmental Carcinogens,* prepared by Max R. Peterson and Dennis F. Naugle of Research Triangle Institute and Michael A. Berry of the U.S. Environmental Protection Agency Environmental

Assessment and Criteria Office, for the U.S.
Environmental Protection Agency, June 1990.

The table of contents is organized in such a way as to serve as a subject index
and provides easy access to the information contained in the book.

Advanced composition and production methods developed by
Noyes Data Corporation are employed to bring this durably
bound book to you in a minimum of time. Special techniques
are used to close the gap between "manuscript" and "completed
book." In order to keep the price of the book to a reasonable
level, it has been partially reproduced by photo-offset directly
from the original reports and the cost saving passed on to the
reader. Due to this method of publishing, certain portions of the
book may be less legible than desired.

NOTICE

Parts I and II of this book were prepared as accounts of work sponsored by the U.S. Department of Energy. Neither the United States Government nor any agency thereof, nor Battelle Memorial Institute, nor any of their employees, nor the Publisher, makes any warranty, expressed or implied, or assumes any legal liability or responsibility for the accuracy, completeness, or usefulness of any information, apparatus, product, or process disclosed, or represents that its use would not infringe privately owned rights. Reference herein to any specific commercial product, process, or service by trade name, trademark, manufacturer, or otherwise, does not necessarily constitute or imply its endorsement, recommendation, or favoring by the United States Government or any agency thereof, Battelle Memorial Institute, or the Publisher. The views and opinions of authors expressed herein do not necessarily state or reflect those of the United States Government or any agency thereof or the Publisher.

Part III of the book was prepared as an account of work sponsored by the U.S. Environmental Protection Agency. It was reviewed in accordance with USEPA policy and approved for publication. On this basis the Publisher assumes no responsibility nor liability for errors or any consequences arising from the use of the information contained herein. Mention of trade names or commercial products does not constitute endorsement or recommendation for use by the Agency or the Publisher.

The book is intended for information purposes only. The reader is cautioned to obtain expert advice before implementation of any procedures suggested in the book for mitigating sick building syndrome, lest any of them be hazardous. Final determination of the suitability of any information or procedure for use by any user, and the manner of that use, is the sole responsibility of the user.

All information pertaining to law and regulations is provided for background only. The reader must contact the appropriate legal sources and regulatory authorities for up-to-date regulatory requirements, and their interpretation and implementation.

Contents and Subject Index

PART II
HEALTH EFFECTS

PART III
SUGGESTED METHODS OF ANALYSIS FOR
INDOOR AIR ENVIRONMENTAL CARCINOGENS

Part I

Sources, Mitigation

The information in Part I is from *Indoor Air Quality Issues Related to the Acquisition of Conservation in Commercial Buildings,* prepared by M.C. Baechler, D.L. Hadley, and T.J. Marseille of Pacific Northwest Laboratory (operated by Battelle Memorial Institute for the U.S. Department of Energy) for Bonneville Power Administration under an agreement with the U.S. Department of Energy, September 1990.

DISCLAIMER

Summary

The quality of indoor air in commercial buildings is dependent on the complex interaction between sources of indoor pollutants, environmental factors within buildings such as temperature and humidity, the removal of air pollutants by air-cleaning devices, and the removal and dilution of pollutants from outside air. To the extent that energy conservation measures (ECMs) may affect a number of these factors, the relationship between ECMs and indoor air quality is difficult to predict. For example, depending on the location of pollutant sources (indoor or outdoor), indoor pollutant concentrations may be either increased or decreased by a reduction in ventilation. For instance, additional caulking and weatherstripping will reduce the infiltration/ exfiltration rate. This reduction in ventilation may increase indoor pollutant levels originating from an indoor source because of a reduction in fresh air dilution and removal. On the other hand, the same ECM will decrease the levels for a pollutant with an outdoor source location by reducing the transport of that pollutant to the indoor environment.

Energy conservation measures may affect pollutant levels in other ways. Conservation measures, such as caulking and insulation, may introduce sources of indoor pollutants. Measures that reduce mechanical ventilation may allow pollutants to build up inside structures. Finally, heating, ventilating, and air-conditioning (HVAC) systems may provide surface areas for the growth of biogenic agents, or may encourage the dissemination of pollutants throughout a building.

Conservation measures that involve carefully maintaining and operating mechanical equipment may improve the quality of indoor air. In some instances, proper operation and design will increase ventilation quality and effectiveness and thus improve pollutant removal mechanisms. In addition, inspections and proper installation will make the buildup of biogenic colonies less likely in HVAC systems.

Information about indoor air quality and ventilation in both new and existing commercial buildings is summarized in this report. Sick building syndrome and specific pollutants are discussed, as are broader issues such as ventilation, general mitigation techniques, and the interaction between

2

energy conservation activities and indoor air quality. Pacific Northwest
Laboratory[a] (PNL) prepared this review to aid the Bonneville Power
Administration (Bonneville) in its assessment of potential environmental
effects resulting from conservation activities in commercial buildings.

---

[a] Pacific Northwest Laboratory is operated by Battelle Memorial Institute
     for the U.S. Department of Energy under Contract DE-AC06-76RLO 1830.

1. Introduction

The Bonneville Power Administration (Bonneville) has taken a leading role among federal agencies in assessing the environmental impacts of indoor air pollutants and designing appropriate program responses. These efforts have included extensive research programs into residential ventilation and indoor pollution characterization and monitoring. Bonneville has also prepared three environmental documents under the National Environmental Policy Act (NEPA) that focused on indoor air quality issues (Bonneville 1982, 1984, 1988).

One of these documents, the 1982 *Environmental Assessment of Energy Conservation Opportunities in Commercial-Sector Facilities in the Pacific Northwest* (Bonneville 1982), supported conservation programs in existing commercial buildings. Bonneville is now planning the implementation of aggressive commercial conservation acquisition programs in both new and existing buildings. Because of changing information, and because of the change in scope of the programs being designed, Bonneville is now reexamining the potential environmental effects of conservation activities in commercial buildings. Pacific Northwest Laboratory[a] (PNL) prepared this review of information about indoor air quality in commercial buildings to aid Bonneville in its assessment of potential environmental effects.

This report summarizes information about indoor air quality and ventilation in both new and existing commercial buildings. The quality of indoor air is dependent on the complex interaction between sources of indoor pollutants, environmental factors within buildings such as temperature and humidity, the removal of air pollutants by air cleaning devices, and the removal and dilution of pollutants from inside air by ventilation. This report addresses these issues for specific pollutants in Section 2, and more generally in Sections 3 and 4. Section 5 focuses on linking indoor air quality issues with specific conservation measures.

(a) Pacific Northwest Laboratory is operated by Battelle Memorial Institute for the U.S. Department of Energy under Contract DE-AC06-76RLO 1830.

4

2. Pollutant Characterization

The dynamics of specific types of pollutants in the commercial building environment are reviewed in this section. Specific pollutant types include volatile organic compounds (VOC), respirable suspended particulates (RSP), biological contaminants, combustion gases, and radon. Before discussing pollutants individually, Section 2.1 addresses the characteristics of pollutants and the sick building syndrome.

2.1 SICK BUILDING SYNDROME AND THE COMBINED EFFECTS OF POLLUTANTS

One national survey reports that 25% of American workers feel that the quality of their workplace air affects their work adversely (Sheldon et al. 1988a). Woods (as reported in Levin 1989a) found that 20% of the office workers in the United States may be affected by sick building syndrome (SBS). Sick building syndrome refers to health and comfort problems associated with working or being in a particular building (EPA 1988). The term generally applies to problems related to indoor air pollution, rather than describing complaints stemming solely from humidity control or inadequate temperature. If the problems result in clinically defined illness, disease, or infirmity in occupants, the building is said to manifest building-related illness (EPA 1988). These conditions may be caused by a number of interacting factors. The Environmental Protection Agency (EPA) lists the following key contributing factors:

- inadequate ventilation
- pollutants emitted inside of buildings
- contamination from outside sources
- biological contamination.

These factors interact with other environmental considerations such as inadequate temperature, uncomfortable humidity, or poor lighting. The causes of SBS and building-related illness are complex. Once a building has been implicated, the causes often remain undetermined. Perhaps the best explanation for occupant responses in the absence of standardized testing and pollutant concentration requirements was offered by Carey et al. (1985):

5

"Virtually all cases of sick building syndrome are discovered by nature's imperfect pollution detector, the human body."

Molhave (as reported in Levin 1989a) reports the following common symptoms reported in cases of SBS:

- sensoric irritation in eye, nose or throat, such as dryness, stinging, smarting, irritating sensation, hoarseness, and changed voice

- skin irritation, such as reddening of skin, stinging, smarting, itching sensation, dry skin

- neurotoxic symptoms such as mental fatigue, reduced memory, lethargy, drowsiness, reduced power of concentration, reduced memory, headache, dizziness, intoxication, nausea, tiredness

- unspecified hyperreactions, such as running nose and eyes, asthma-like symptoms, respiratory sounds

- odor and taste complaints, such as changed sensitivity and unpleasant odor or taste.

Molhave (1990) further suggests that human response may be related to a biological model consisting of three stages:

- sensory perception of the environment - The senses include odor, taste, and chemical sense. Chemical sense refers to nerves in the mucousal membranes and skin that react to chemical stimuli. Activation of the senses leads to irritation and possibly a protective response, such as sneezing.

- weak inflammatory reactions - Inflammatory reactions are related to microbiologic, metabolic, or immune system reactions and are generally considered to be a protective reaction to potential cell damage. Acute, reversible reactions seem to be relevant to low-level pollutant exposure in nonindustrial environments.

- environmental stress reactions - The constant effort needed to identify wanted and to override unwanted sensory information, and the efforts needed to maintain protective reflexes may cause secondary effects, such as headache.

Key pollutants that may be associated with SBS are discussed in Sections 2.2 through 2.6. Although it is not the source of the pollutants causing distress, inadequate ventilation is an often-cited cause of sick buildings (Burge 1990). As of December 1985, the National Institute for Occupational

Safety and Health (NIOSH) had conducted 365 health-hazard evaluations of building-related problems. The NIOSH has concluded in most cases that the problem could be eliminated by improved ventilation (Melius et al. 1984).

Burge (1990) states that researchers consistently link an increased number of symptomatic building occupants with microbiological contamination either from dampness or from chillers and humidifiers. However, when airborne levels of bacteria and fungi are measured, no direct relationship has been found with SBS symptoms and airborne levels. It is likely that the problems relate to mycotoxins and endotoxins, which microbiological organisms release. There is also increasing evidence that biocides, such as isothialzolones, glutaraldehyde, and chlorhexidine in low concentrations may cause symptoms of SBS.

Recently Molhave (1990) has postulated that total volatile organic compounds (TVOC) can be used to define a concentration level at which SBS symptoms begin. The no effect level seems to be at 0.16 mg/m^3. An exposure range of from 0.2 to 3 mg/m^3 may result in possible health effects. At concentrations greater than 3 mg/m^3, discomfort is expected; at concentrations greater than 25 mg/m^3, toxic effects may appear. However, the TVOC must be grouped using common characteristics. At present there is little agreement on how these groupings should be established. Siefert (1990) points out that until conventions are established for these groupings, it will be difficult to promulgate regulations or guidelines.

Principles for a future ventilation standard that acknowledges all pollution sources have been proposed by Fanger (1990, 1989). Required ventilation levels would be calculated based on the total indoor pollution load, the available outdoor air quality, and the desired level of indoor air quality. In this standard two new units would be utilized, the olf and decipol. Perceived air quality is expressed in decipol. One decipol is the perceived air quality in a space with a pollution strength of one olf, ventilated by 10 l/s of clean air (1 decipol - 0.1 olf/[l/s]). One olf is defined as the pollution from a standard person. However, any pollution source can be expressed in olfs. The conversion would be the number of standard persons required to make the air as annoying as the actual pollution source. Panels of judges would make subjective measurements of the level of annoyance. Thus, given a certain percentage of dissatisfied among a panel of judges, the number of standard persons (olfs) that would cause the same dissatisfaction could be calculated. The olf value is not related to possible health risks of pollutants.

Other investigations have also found a correlation between SBS and ventilation conditions. Levin (1989a) reports that four British studies found higher symptom prevalence rates in sealed, mechanically ventilated buildings than in those naturally ventilated. One of these studies found that buildings with humidification had the greatest symptom rates. A third study found a threefold excess of symptom rates in mechanically ventilated buildings compared to naturally ventilated buildings. A study of 30 British buildings found little difference between natural and mechanically ventilated buildings, but did find higher symptom prevalence in air-conditioned buildings as compared to non-air-conditioned buildings. This study also found higher symptom rates in women than in men, and in public-sector buildings than in private-sector buildings.

In contrast to these studies finding ventilation to be a significant causal factor, several investigators have suggested that multiple factors contribute to SBS, none of which alone would result in complaints and symptoms (Levin 1989a). A comprehensive study known as the Danish Town Hall Study investigated 27 buildings and found that differences in the prevalence of symptoms among buildings was significant and was correlated with building characteristics and occupant factors. Building and occupational factors that explained differences in symptom prevalence included the following:

- Age of the building, wherein the lowest prevalence of symptoms was found in the oldest buildings. Symptoms varied greatly by building, suggesting that the symptoms were building-related.

- Elevated rates of mucosal irritation were associated with the total weight and potentially allergenic fraction of floor dust, the presence of open shelves, the area of fleecy material, the number of work stations, and air temperature.

- Symptoms correlated strongly with job category, with the highest prevalence of symptoms found in subordinate job positions. Photo-printing, working at video display terminals, and handling carbon-less paper was related to both mucosal and general symptoms.

- Women had a higher symptom rate than men and complained more frequently about indoor climate. The number of weekly working hours of women weakly corresponded with reports of mucosal and general symptoms.

Lighting can also affect occupants' perceptions of buildings. Sterling and Sterling (1984) report that there is a significant relationship between poor lighting and reported building illness. Office workers having poor building lighting were found to be more likely to think of their buildings as contributing to poor health. Noise has also been identified as a potential cause of SBS symptoms.

The importance of indoor air quality is being recognized with an increase in legislative attention. In the spring of 1989, the Indoor Air Quality Act of 1989 (H.R. 1530 and S. 657) was introduced in both houses of the U.S. Congress. Though the bill has yet to be passed, it has 47 current cosponsors in the House and 19 cosponsors in the Senate.[a] The bill directs the Administrator of the EPA to develop a national research, development, and demonstration program. This program would focus on human health effects and the identification of types and levels of contaminants. Other features would include the funding of demonstration projects, the formation of a national clearinghouse, and the establishment of advisory pollutant levels.

2.2 VOLATILE ORGANIC COMPOUNDS

Volatile organic compounds refers to a class of carbon-based chemicals that evaporate easily at room temperature and thus give off vapors that can be inhaled.

(a) Personal communication with Gordon J. Evans, District Assistant to Congressman Ron Wyden. 10 August 1990.

2.2.1   Volatile Organic Compound Sources

The explosion of new building materials, consumer goods, and office equipment developed since World War II has made VOC sources ubiquitous. Wallace (1987) has concluded that nearly every home and place of business contains common materials that may cause elevated levels of chemical exposure. Over 900 separate volatile compounds have been found in indoor air. Organic compounds can be grouped according to structural similarities. Examples of organic compound types and their sources are shown in Table 2.1.

Wallace identified from 10 to 82 vapor compounds (stainless steel polish had the most) in chamber study experiments. The study and experiments assessed 15 materials grouped in the four categories of cleaners and polishes, floor and wall coverings, glues, and pesticides.

In other chamber studies, the National Aeronautics and Space Administration (NASA) has tested over 5000 materials that may find their way onto space craft. Ozkaynak (1987) has divided these materials into groupings such as adhesives, coatings, fabric, foam, labels, lubricant, papers, plastic, rubber, tape, thread, velcro, wire, clothing, computer equipment, cosmetics, deodorants, electrical equipment, and so on. Ozkaynak reported findings for 19 VOCs. Of these, benzene was found in 387 materials, styrene in 284, trichloroethylene in 381, and ethyl benzene in 19. The highest concentrations were found in health and beauty aids. Mass balance calculations of indoor pollutant concentrations have also shown paint to be a substantial source (Baechler et al. 1989).

Sheldon et al. (1988b) found that building materials collected from new buildings emitted high rates of the compounds found in the buildings. Further, these compounds were similar in two new buildings, suggesting that the findings may be generally true for many buildings. Materials emitting these compounds at the highest rates were surface coatings such as adhesives, caulking, and paints; wall and floor coverings such as molding, linoleum tile, and carpet; and other materials such as telephone cable.

TABLE 2.1. Examples of Organic Compounds and Sources

Pollutant Type	Organic Compounds	Indoor Sources
Aliphatic and oxygenated aliphatic hydrocarbons	α-pinene, n-decane, n-undecane, n-dodecane, propane, butane, n-butylacetate, ethoxy ethyl acetate octane	Cooking and heating fuels, aerosols propellants, cleaning compounds, paints, carpet, moldings, particle board, refrigerants, lubricants, flavoring agents, perfume base
Halogenated hydrocarbons	Chloroform, methyl chloroform, dichloromethane, polychlorinated biphenyls, 1,1,1-trichloroethane, trichloroethylene tetrachloroethylene, chlorobenzene, dichlorobenzene carbon tetrachloride	Aerosol propellants, fumigants, pesticides, refrigerants, adhesives, caulk, paint, linoleum tile, carpet, latex paint, and degreasing, dewaxing, and dry cleaning solvents
Aromatic hydrocarbons	Xylene, ethyl benzene, trimethylbenzene, ethyltoluene, propylbenzene, benzene, styrene, toluene	Paints, varnishes, glues, enamels, lacquers, cleaners, adhesives, molding, insulation, linoleum tile, carpet
Alcohols	Ethanol, methanol	Lacquers, varnishes, polish removers, adhesives
Ketones	Acetone, diethyl ketone, methyl ethyl ketone	Lacquers, varnishes, polish removers, adhesives, cleaners
Aldehydes	Formaldehyde, nonanal	Fungicides, germicides, disinfectants, artificial and permanent-press textiles, paper, particle boards, cosmetics, flavoring agents

2.2.2 Volatile Organic Compound Source Strengths

Emission rates from 31 materials collected from new office buildings are shown in Table 2.2. The results show that while some solvent-based materials were the highest emitters, other solvent-based building materials emitted none of the target compounds (Sheldon et al. 1988b).

Building materials, such as caulk and insulation, associated with energy-efficiency measures have been found to emit VOCs. Table 2.3 summarizes emissions from the materials that have been tested in chamber studies. These emissions and their rates can be compared with those listed in Table 2.2. An environmental assessment (EA) prepared for the U.S. Department of Energy (DOE) (1986) concluded that emissions from five insulation types did not contribute significantly to indoor VOC concentrations. This conclusion was based on mass-balance calculations using emission rates from chamber study tests of building materials. Table 2.3 summarizes chamber studies of building materials associated with energy-efficient construction and covers a broader range of materials than those included in the EA.

2.2.3 Commercial Building Volatile Organic Compound Concentrations

The levels of individual VOCs found in buildings are often several times below threshold limit values (TLVs) for occupational settings, or levels considered to be harmful for any one chemical in an occupational setting. However, many indoor VOC concentrations have been found to be much higher than levels found outdoors. Sheldon et al. (1988a) found indoor-outdoor ratios of total organics of 2 or 3 to 1 in three older buildings. In a new office building, this ratio was 50:1, dropping to 10:1 after two months, and 5:1 after three additional months. A total of about 500 compounds were found at least once from all of the buildings sampled.

In a companion study, Sheldon et al. (1988b) found indoor levels of total organics in two new buildings up to 400 times greater than outdoor concentrations. After several months these concentrations dropped to 3 to 30 times outdoor levels. Table 2.4 gives a comparison of mean VOC concentrations

TABLE 2.2. Summary of Emission Results from Building Materials[a]

Sample	Aliphatic and Oxygenated Aliphatic Hydrocarbons	Aromatic Hydrocarbons	Halogenated Hydrocarbons	All Target Compounds
Cove adhesive	(b)	(b)	(b)	>5000
Latex caulk	252	380	5.2	637
Latex paint A	111	52	86	249
Carpet adhesive	136	98	(c)	234
Black rubber molding	24	78	0.88	103
Small-diameter telephone cable	33	26	1.4	60
Vinyl cove molding	31	14	0.62	46
Linoleum tile	6.0	35	4.0	45
Large-diameter telephone cable	14	20	4.3	38
Carpet	27	9.4	---	36
Vinyl edge molding	18	12	0.41	30
Particle board	27	1.1	0.14	28
Polystyrene foam insulation	0.19	20	1.4	22
Tar paper	3.2	3.1	---	6.3
Primer/adhesive	3.6	2.5	---	6.1
Latex paint B	---	3.2	---	3.2
Water repellent mineral board	1.1	0.43	---	1.5
Cement block	---	0.39	0.15	0.54
Polyvinyl chloride pipe	---	0.53	---	0.53
Duct insulation	0.13	0.15	---	0.28
Treated metal roofing	---	0.19	0.06	0.25
Urethane sealant	---	0.13	---	0.13
Fiberglass insulation	---	0.08	---	0.80
Exterior mineral board	---	0.03	---	0.03

(a) Source: Adapted from Sheldon et al. (1988b).
(b) Emission rate for cove adhesive is a minimum value; sample was
     overloaded. It is estimated that cove adhesive is one of the highest
     emitters of volatile organics.
(c) No detectable emissions.

TABLE 2.3.   Summary of Emissions from Energy-Efficient Building Materials

Investigator	Technique	Source	Emissions	Concentration or Emissions Rate
Miele 1989	Large scale test chamber	R-30 bonded	Formaldehyde--this was the only compound measured--reported values are chamber concentrations.	0.065 ppm @ 77°F [a]
		Fiberglass insulation Kraft-faced		0.072 ppm @ 77°F
		R-19 Kraft-faced		0.072 ppm @ 77°F
		R-11 Kraft-faced		0.072 ppm @ 77°F
		R-11 unfaced		0.085 ppm @ 77°F
		R-13 unfaced		0.085 ppm @ 77°F
		Duct Line--non-woven fabric facing		0.015 ppm @ 77°F [b]
		Duct Line--non-woven fabric facing		0.020 ppm @ 110°F [b]
		Duct Board--1" foil scrim Kraft facing		0.020 ppm @ 77°F [b]
		Duct Board--1" foil scrim Kraft facing		0.028 ppm @ 77°F [b]
		Duct Board--1" foil scrim Kraft facing		0.051 ppm @ 110°F
				0.070 ppm @ 110°F
Tichenor and Mason 1988	Test chamber	Silicone caulk	Total organics--major compounds identified include: methyl ethyl ketone, butyl propionate, 2-butoxyethanol, butanol, benzene, and toluene. Sample age: 0.5 hr Sample age: 1.0 hr Sample age: 5.0 hr	50.0 mg/m²-h 30.0 mg/m²-h 2.0 mg/m²-h
White, Reaves, Fleist, and Mann 1988	Review of test chamber data	Caulking compound	C-10 alkene C 4-alcohol	1.23 mg/m²-h <0.01 mg/m²-h
			N-decane	6.82 mg/m²-h
			N-undecane	<0.01 mg/m²-h
			Toluene	0.02 mg/m²-h
			Ethyl benzene	7.32 mg/m²-h
			2-xylene	1.02 mg/m²-h
			3-xylene	2.77 mg/m²-h
			4-xylene	24.20 mg/m²-h
		Insulation foam	VOC	2.3 mg/m²-h

TABLE 2.3.   (contd)

Investigator	Technique	Source	Emissions	Concentration or Emissions Rate
van der Wal, Mouns, and Steelage 1987	Test chamber	Rock wool and glass fibers. The following compounds were found in at least one sample of either insulation type taken from a home, school, or office.	18 aldehydes and ketones	Not qualified-- concentration greatest when samples were wet.
Sheldon, Thomas and Jungers 1986	Test chamber	Fiberglass insulation	Aromatic hydrocarbons	0.00008 mg/m^2-h
		Polystyrene foam and insulation	Aromatic hydrocarbons	0.020 mg/m^2-h
			Halogenated organics	0.00134 mg/m^2-h
			Volatile aromatic compounds	0.02017 mg/m^2-h
			Aliphatic and oxygenated aliphatic compounds	0.00019 mg/m^2-h
			2-butene-1-ol, Dentane, 1,2-dimethyl-cyclopropane, benzonitrile, benzaldehyde	Not quantified
		Duct insulation	Aromatic hydrocarbons	0.00015 mg/m^2-h
			Aliphatic and oxygenated aliphatic organics	0.00013 mg/m^2-h
			Trimethylhexene	Not quantified[c]
		Latex caulk	Halogenated organics	0.005[c] mg/m^2-h
			Volatile aromatic compounds	0.3809[c] mg/m^2-h
			Aliphatic and oxygenated aliphatic organics	0.251[c] mg/m^2-h
			Aciphatic hydrocarbons, aromatic hydrocarbons	Not quantified
Molhave 1982	Test chamber	Silicone caulk	23 compounds	26.0 mg/m^2-h
		Mineral wool	13 compounds	0.012 mg/m^2-h
		Polystyrene foam insulation	15 compounds	1.4 mg/m^2-h
		Polyurethane foam insulation	5 compounds	0.12 mg/m^2-h
Miksch, Hollowell, and Schmidt 1982	Test chamber	Phenoseal Brand caulk[d]	Butylacetate, 2-ethanol	Not quantified
		Vinyl acrylic latex caulk[d]	Ethylene glycol	Not quantified
		Butyl caulk[d]	Octane, octene, nonane, aliphatic hydrocarbons, alkylbenzenes	Not quantified

(a)   Chamber background formaldehyde levels ranged from 0.024 ppm to 0.028 ppm.
(b)   Formaldehyde concentration after test was less than background level, suggesting test material was a sink.
(c)   Minimum value, compound saturated the detector during analysis.
(d)   Sample aged 24 hours before vapor collected.

TABLE 2.4.   Mean Indoor Concentration of Volatile Organic Compounds Found in Public Buildings

	Martinsburg, WV Hospital (New)[a]			Fairfax, VA Office (New)[b]		Worcester, MA Nursing Home (New)[c]		Washington, D.C. Office (Old)	Cambridge, MA Office/School (Old)	Martinsburg, WV Nursing Home (Old)
	Mean Indoor Concentration (mg/l)									
Compounds	Trip 1 (7/84)	Trip 2 (10/84)	Trip 3 (8/85)	Trip 1 (1/85)	Trip 2 (4/85)	Trip 1 (4/85)	Trip 2 (8/85)	Trip 1 (8/84)	Trip 1 (2/85)	Trip 1 (7/84)
Aromatic Hydrocarbons										
Benzene	1.55	2.13	2.88	2.74	4.95	1.70	2.44	5.61	4.50	3.13
m-Xylene	6.88	3.13	9.91	41.53	15.05	23.80	5.33	27.11	8.72	2.95
o-Xylene	3.05	0.91	3.07	18.40	3.67	8.92	2.07	9.28	3.43	0.99
Styrene	1.00	1.07	1.33	2.52	2.87	2.99	1.27	2.36	1.32	1.19
Ethylbenzene	1.94	1.01[d]	2.88	51.26	5.37	7.90	2.15	10.15	2.69	0.97
Isopropylbenzene	0.31	ND[d]	0.33	3.94	0.67	2.27	0.33	0.79	0.36	ND
n-propylbenzene	ND	ND	ND	5.00	1.13	2.99	0.70	1.22	0.56	ND
m-Ethyltoluene	1.11	0.86	1.48	27.41	5.57	12.38	2.62	6.07	2.62	0.90
o-Ethyltoluene	ND	ND	0.66	8.89	2.08	4.01	0.73	1.60	0.74	ND
1,2,3-Trimethylbenzene	0.63	0.43	0.76	15.10	2.91	5.32	0.72	1.80	1.06	0.79
1,2,4-Trimethylbenzene	1.48	0.98	1.82	73.51	7.27	13.95	2.52	6.28	2.80	0.98
1,2,5-Trimethylbenzene	ND	ND	0.75	16.97	2.75	6.83	0.92	1.83	1.14	ND
Aliphatic Hydrocarbons										
a-Pinene	ND	ND	ND	14.13	24.64	5.19	ND	ND	2.65	ND
n-Decane	3.65	2.73	2.71	436.38	15.24	68.27	3.81	2.26	5.98	1.67
n-Undecane	3.31	1.96	2.34	210.80	33.93	68.51	3.48	2.85	6.77	ND
n-Dodecane	ND	ND	ND	152.69	23.74	31.42	ND	ND	2.23	ND
Chlorinated Hydrocarbons										
1,2-Dichloroethane	2.06	1.49	2.21	ND	4.51	ND	ND	ND	ND	ND
1,1,1-Trichloroethane	4.98	4.50	15.54	12.54	38.85	4.03	1.76	40.98	10.69	3.09
Trichloroethylene	1.05	ND	ND	ND	7.93	2.58	0.57	0.61	10.89	ND
Tetrachloroethylene	ND	ND	1.79	ND	1.64	1.13	0.96	3.97	4.11	0.99
p-Dichlorobenzene	ND	ND	6.61	ND	2.64	2.17	0.62	0.60	ND	ND
Oxygenated Hydrocarbons										
n-Butylacetate	ND	ND	ND	ND	6.34	ND	1.22	2.63	1.48	ND
2-Ethoxyethyl acetate	1.31	ND	ND	ND	2.16	9.58	ND	1.67	ND	ND

(a) Building completed -34 weeks before first monitoring trip.
(b) Building completed -1 week before first monitoring trip.
(c) Building completed -4 weeks before first monitoring trip.
(d) Below the quantifiable limit.
Source adapted from Shelton et. al. 1988b

found in a variety of building types, from a paper by Jungers and Sheldon (1987). VOC concentrations in three new buildings measured soon after construction and after occupancy are shown in Table 2.5 (Wallace et al. 1987).

In 109 total VOC samples taken from 15 office buildings where occupants reported widespread occurrence of annoyance and discomfort on at least one floor or zone, Morey and Jenkins (1989) found that 20% of the samples exceeded or were equal to 1000 $\mu g/m^3$. And 6% of the buildings had concentrations equal to or exceeding 2000 $\mu g/m^3$. Table 2.6 shows concentrations of total VOCs from the 15 buildings.

In a study of 38 commercial buildings in the Pacific Northwest, Turk et al. (1987) found low formaldehyde concentrations. The average levels in 21 of the buildings were above the 20-parts per billion (ppb) detection limit. None of the buildings had whole-building mean levels exceeding the 1981 ASHRAE guideline of 100 ppb. Three buildings had comparatively elevated levels. The researchers suggest that the elevated levels in one building were caused by formaldehyde-bonded products used in new construction and, in another building, similar products used in remodeling. The likely source in a third building was a print shop that shared the same structure.

Table 2.7 is adapted from the 1986 U.S. Department of Energy EA and shows calculated organic concentrations for commercial buildings in Seattle.

For comparison, Molhave (1985) has suggested that people begin sensing total VOCs in indoor air at about 1000 $\mu g/m^3$. Molhave's work also suggests that SBS may result from the additive or synergistic effect of the complex mixture of VOCs found in buildings, rather than from any single pollutant (Molhave 1985; Molhave et al. 1985). His work may be among the first objective proof that a major component of the SBS may be common indoor organic vapors at levels far below occupational health standards (Wallace 1985).

2.2.4 Environmental Factors

Organic vapors are released to indoor air from a multitude of products, cleaning agents, and building materials. The factors affecting the release and interaction of these vapors are complex and include temperature, humidity,

TABLE 2.5.   Mean 3-Day Concentrations ($\mu g/m^3$) of Organics in Three New Buildings

	Recently Constructed			Following Occupancy				Maximum Outdoor Concentration
	Office 1	Office 2	Nursing Home	Office 1	Office 2	Office 1	Nursing Home	
Number of Samples	20	18	18	19	18	20	18	6
Time Since Completion (Weeks)	1	2	6	7	15	22	23	--
Aliphatics								
α-Pinene	NM(a)	14	5	NM	25	NM	ND(b)	ND
Decane	380	440	68	38	15	4	4	4
Undecane	170	210	69	48	34	13	4	2
Dodecane	47	150	31	19	24	5	ND	1
Aromatics								
Benzene	5	3	2	7	5	7	2	5
Xylenes	214	59	33	27	19	12	7	6
Styrene	8	3	3	7	3	4	1	4
Ethylbenzene	84	51	8	6	5	5	2	2
Propylbenzenes	NM	9	5	NM	2	NM	1	ND
Ethyltoluenes	NM	36	16	NM	8	NM	3	2
Trimethylbenzenes	NM	110	26	NM	13	NM	4	3
Chlorinated								
1,2-Dichloroethane	NM	ND	ND	NM	5	NM	ND	ND
1,1,1-Trichloroethane	380	13	4	100	39	49	2	10
Trichloroethylene	1	ND	3	38	8	27	1	1
Tetrachloroethylene	7	ND	1	2	2	3	1	1
Dichlorobenzenes	1	ND	2	1	3	1	1	ND
Chloroform	1	NM	NM	2	NM	18	NM	17
Carbon tetrachloride	1	NM	NM	1	NM	1	NM	1
1,2-Dichloroethane	NM	ND	ND	NM	5	NM	ND	ND

(a) Not measured.
(b) Not detected.
Source is adapted from Wallace et al 1987

TABLE 2.6. Concentrations of Total Volatile Organic Compounds in 15 Buildings

Location	Concentration of TVOC in $\mu g/m^3$	Number Samples	Percent of Total Samples
Indoors	=> 2000	6	6
Indoors	=> 1000 to 1999	15	14
Indoors	=> 500 to 999	25	23
Indoors	=> 200 to 499	31	28
Indoors	=> 100 to 199	19	17
Indoors	=> 1 to 99	13	12
Outdoors	=> 500 to 999	3	13
Outdoors	=> 200 to 499	5	22
Outdoors	=> 100 to 199	7	30
Outdoors	=> 1 to 99	8	35

Average of 109 indoor samples = 660 $\mu g/m^3$
Average of 23 outdoor samples = 232 $\mu g/m^3$
Source: Adapted from Morey and Jenkins 1989

TABLE 2.7. Projected Total Organic Concentration Increments in Buildings with Unaged Building Materials (mg/m^3)

Building Category	ASHRAE Standard 90A-1980 Design	Proposed Standard Design	Increment
Small Office	0.082	0.594	0.511
Medium Office	0.240	0.432	0.191
Large Office	1.005	0.006	-0.999
Retail Store	0.013	0.199	0.185
Strip Store	0.002	0.003	0.002
Apartment	0.315	0.792	0.478
Hotel	0.031	0.043	0.013
Warehouse	1.217	16.194	14.976
Assembly	0.001	0.010	0.009
School	0.020	0.024	0.004

time, ventilation, source quantity, the volume of the affected space, the age of the material, and the tendency of walls and other surfaces to absorb and re-emit the pollutant.

Perhaps the best studied of the organics is formaldehyde. Hawthorne and Mathews (1985) have developed models of formaldehyde emissions from pressed wood products using urea-formaldehyde-based glue. These models show that emission rates increase when air exchange rates increase, product loads decrease, or concentrations decrease. Researchers have also demonstrated that formaldehyde emissions tend to increase as humidity or temperature levels increase. Hawthorne and Mathews (1985) suggest that the expected length of time for a given source of formaldehyde to release about half of its emissions (half-life) may be on the order of a few years. Spengler and Sexton (1983) report a half-life of 4.4 years.

As described earlier in this section, organic emissions testing has been done on materials other than formaldehyde. However, it is difficult to evaluate the limited data on organic emissions because standard testing protocols have not been developed (Tichenor and Mason 1988; Baechler et al. 1989).

For wet materials such as caulk, Tichenor and Mason (1988) have con-cluded that time, or the age of the sample, is a critical parameter; source durations can be as short as a few hours. Figure 2.1 shows the emission factor of total measured organics from silicone caulk over time at two air exchange rates. Emission rates decrease rapidly with time as the VOCs are depleted from the source. The emission rates are initially higher with the

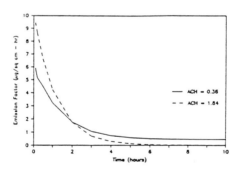

FIGURE 2.1.   Caulking Compound Emission Factor Versus
              Time Total Measured Organics

high air exchange rate, but after 2 hours the emission rates are higher for
the low air exchange rate. The high air exchange rate caused the VOCs to be
emitted more quickly.

Time can also play an important role in VOC sources that are not wet.
As discussed in Section 2.2.2, Sheldon et al. (1988a, 1988b) found substantial
reductions in indoor VOC concentrations as a building ages. According to
Sheldon et al. (1988a), the half-lives of the compounds found in two new
office buildings ranged from 2 to 8 weeks. Thus, based on this limited data,
Sheldon concludes that the time for the building to approach outdoor VOC
concentrations would range from 3 to 12 months. As time passes, the original
source may be depleted. But sinks--sources that adsorb the pollutant--can re-
emit the pollutant to the air (Tichenor et al. 1990; Berglund, et al. 1988).
Modeling demonstrates that fast-emitting sources could be depleted within
2 days, but re-emission from sinks can affect indoor concentrations for up to
2 weeks (Tichenor et al. 1990).

Tichenor and Mason (1988) also found that a mixture of organic compounds
may be emitted from a single source at different rates. Figure 2.2 shows
emission factors over time for three of the compounds emitted from silicone
caulk. The emission rates vary at any given point in time and decrease at
different rates.

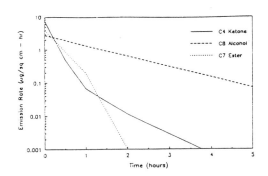

FIGURE 2.2.   Caulk Emissions Versus Time -- Three
                Compounds. T = 23°C; RH = 50%, ach = 1.84

2.2.5 Standards, Guidelines, and Regulatory Effects

Indoor air quality typically is not regulated in nonoccupational settings. Table 2.8 lists the consensus findings of a World Health Organization
(WHO) working group on indoor pollutant concentrations. Table 2.9 shows U.S.
standards and guidelines for formaldehyde. And Table 2.10 shows occupational
threshold limit values (TLVs) and short-term exposure limits (STELs) for a
small sample of organics commonly found in indoor air.

Product standards provide one means of limiting the use of high polluting sources. An example of this type of measure is the formaldehyde product
standard that Bonneville now enforces in the residential new homes programs.
The standard requires that structural building materials used in Bonneville's
new homes programs meet the U.S. Department of Housing and Urban Development's
(HUD) requirements for manufactured housing, or be rated for exterior use.
This product standard helps to reduce formaldehyde emissions from products
using urea-formaldehyde-based glues.

Possible means of regulating VOCs are also discussed in Section 2.1.

2.2.6 Interaction with Energy Conservation Measures

Energy conservation measures may affect VOC concentration in one of two
ways: by introducing a source of pollutants, or by changing the rate at which
pollutants are removed from the indoor space. Potential sources of indoor
VOCs from energy conservation measures include caulking compounds and insulation. These materials are discussed in Sections 2.2.1 and 2.2.4. Of the two
potential sources discussed, caulking emits the highest rates of VOCs, but
these are likely to be quickly dissipated. Insulation materials emit pollutants at a lower rate, but have the potential of emitting them over a longer
period of time. Emissions from fibrous insulation may be related to how moist
the material is (van der Wal 1987).

The second way that energy conservation measures may affect indoor VOC
levels is by changing the rate at which pollutants are removed from a space.
The most common removal mechanism involves the exchange of indoor and outdoor
air in ventilation systems. If conservation measures involve reducing the
quantities of outside air without including air cleaning equipment, indoor

TABLE 2.8.   World Health Organization Working Group Consensus of Concern About Indoor Air Pollutants at 1984 Levels of Knowledge

Pollutant[a]	Concentrations[b] Reported	Concentrations[b] of Limited or No Concern	Concentration[b] of Concern	Remarks
Tobacco smoke (passive smoking)				
Respirable particulates	0.05-0.7	<0.1	>0.15	Japanese standard 0.15 mg/m³
CO	1-1.5	<2	>5	indicator for eye irritation (only from passive smoking)
Nitros-dimethylamine	(1-50) x 10⁻⁶	---	---	Mutagens under investigation for carcinogenicity
NO_2	0.05-1	0.19	>0.32	
CO		2% carbon monoxyhemoglobin (COHb) 1-100	3% COHb <11	99.9%[c]
Radon and Progenys	10-3000 Bq/M³	-0	70 Bq/m³	>30 Continuous exposure
Formaldehyde	0.05-2	<0.06	>0.12	Swedish standard for new houses Long- and short-term
SO_2	0.02-1	<0.5	>1.35	SO_2 alone short-term
CO_2	600-9000	<1600	>12000	Japanese standard 1800 mg/m³
O_3	0.04-0.4	0.05	0.08	For long-term exposure
Asbestos	<10 fibres/m³	-0	10^5 fibre/m³	
Mineral fibres	<10 fibres/m³	---	---	Skin irritation
Organics				
Methylene chloride	0.005-1	---	350	TLV[d]
Trichlormethene	0.0001-0.02	---	260	NIOSH[e] recommendations
			270	TLV
			135	NIOSH recommendations
Tetrachloroethene	0.002-0.05	---	335	TLV
1,4 Dichlorobenzene	0.005-0.1	---	450	TLV
Benzene	0.01-0.04	carcinogen	carcinogen	
Toluene	0.015-0.07	---	375	TLV
m.p.-Xylene	0.01-0.05	---	435	TLV
n-Nonane	0.001-0.03	---	1050	ILO[f] (1980)
n-Decane	0.002-0.04	---	---	
Limonene	0.01-0.1	---	560	TLV for turpentine

(a) All gases were considered on their own without other contaminants.
(b) Typical ranges of concentration given in mg/m³ unless otherwise indicated and for short-term exposures.
(c) According to Environmental Health Criteria No. 4, Geneva World Health Organization, 1977.
(d) TLV (threshold limit values) established by the American Conference of Governmental Industrial Hygienists (1983-1984). The values are for occupation exposures and should be considered as extreme upper limits for nonoccupational population for very short-term exposures.
(e) National Institute for Occupational Safety and Health (NIOSH), USA.
(f) International Labor Organization (ILO).
--- = no meaningful numbers can be given because of insufficient knowledge.

TABLE 2.9. U.S. Formaldehyde Standards

Pollutant	Indoor Standards	Industrial Workplace Outdoor Standards	Standards
Formaldehyde	Federal: 0.4 ppm target ambient level, HUD standard for manufactured homes, achieved through product emission standards of 0.2 and 0.3 ppm(a) (HUD, 24 CFR 3280.308, 1984) (C-14) State: 0.4 ppm standard for indoor exposure (MN statute 144.495, 1985) (C-15)	No federal standard State air quality limits: CT 12.00 $\mu g/m^3$ 8 hr IL 0.0150 $\mu g/m^3$ 1 yr IN 18.000 $\mu g/m^3$ 8 hr MA 0.2000 $\mu g/m^3$ 24 hr NC 300.00 $\mu g/m^3$ 15 min NV 0.0710 $\mu g/m^3$ 8 hr NY 2.0000 $\mu g/m^3$ 1 yr VA 12,999 μgm^3 24 hr (C-8)	OSHA issued a final rule Dec. 4, 1987 (52 FR 46168) lowering a previous standard to the above levels, which was effective on Feb. 2, 1987. Mine Safety and Health Admins. used ACGIH TLVs (C-13)

(a) Emissions must not produce concentrations greater than these. These concentrations are measured using chamber studies.

TABLE 2.10.   Threshold Limit Values of Some Organics
Frequently Encountered in Indoor Air[a]

Chemical	TLV-TWA		TLV-STEL		Source
	mg/m^3	ppm	mg/m^3	ppm	
Acetone	1780	(750)	2375	(1000)	Lacquer solvent
Ammonia	18	(25)	27	(35)	Cleaner
Benzene	30	(10)	75	(25)	Adhesive, spot cleaner, paint remover
Carbon tetrachloride	30	(5)	125	(20)	Spot cleaner, dry cleaner
Chlorine	3	(1)	9	(3)	Cleaner
Methanol	260	(200)	310	(250)	Paint, spot cleaner
Trichloroethane	1900	(350)	2450	(450)	Cleaning fluid
Methylene chloride	350	(100)	1740	(500)	Paint remover
Trichloroethylene	270	(50)	805	(150)	Dry-cleaning agent
Turpentine	560	(100)	840	(150)	Paint, finish
Xylene	435	(100)	655	(150)	Solvent, paint carrier, shoe dye
Toluene	375	(100)	560	(150)	Solvent, paint carrier, dry cleaning

(a)   Data from American Conference of Governmental Industrial Hygienists 1983.
Source: Bonneville (1988)

levels of VOCs are likely to increase. On the other hand, if conservation
measures such as economizers that increase quantities of outside air are
employed, indoor VOC levels are likely to decrease. Morey and Jenkins (1989)
found that in 25 problem office buildings, indoor VOC concentrations were just
slightly above outdoor levels under economizer conditions. A more thorough
discussion of ventilation is contained in Section 3 of this report.

In existing buildings, energy conservation can affect the sources and
removal rates of indoor pollutants through inspections and improved maintenance of the building structure and the heating, ventilation, and air-
conditioning (HVAC) system. Perhaps the most direct positive effects will be
on microbial growth in the HVAC system and on building surfaces. For example,

caulking around windows may eliminate the entry of wind-driven rain, which fosters the growth of molds and fungi (Rask and Lane 1989). Inspections may also reveal microbial growth areas in ductwork near cooling coils, or design deficiencies such as the placement of air intakes near pollutant sources such as garbage storage areas or parking garages (Rask and Lane 1989; Morey 1988).

However, if building inspections result in reductions in ventilation rates, there is likely to be a negative effect on indoor air quality. Field studies suggest that ventilation rates often exceed those specified in ASHRAE Standard 62-1981 (American Society of Heating, Refrigerating and Air Conditioning Engineers, Inc. 1981). Seton, Johnson and Odell, Inc. (1984) conclude that nominal mechanical ventilation rates based on actual occupancy are significantly higher than the design rates listed in Standard 62-1981. Turk et al. (1987) found that in a sample of 40 buildings, on average, overall building air exchange rates ranged from 2 to 8 times the rates recommended for smoking areas in Standard 62-1981. The ventilation rates contained in Standard 62-1981 for smoking areas (as opposed to the ventilation rates listed for nonsmoking areas) are greater than the single ventilation rates listed in ASHRAE Standard 62-1989 (ASHRAE 1989a). Thus, Turk's findings suggest that in existing buildings, overall building air exchange rates exceed both the new ASHRAE Standard 62-1989 and the old ASHRAE Standard 62-1981. If the overall building air exchange rates are reduced to match either of the standards, there is likely to be a negative effect on indoor pollution concentrations.

2.2.7 Specific Mitigation Techniques

Three broad strategies may be employed to reduce VOC concentrations -- source avoidance, source control, and pollutant removal. Each is discussed in the following paragraphs.

Source Avoidance and Substitution - Source avoidance is the most sure method of reducing pollutants. Avoidance of a specific source is generally a one-time measure, requiring no maintenance. A sampling of pollutants that may be emitted by building materials is presented in Section 2.2.2. Avoidance can be achieved through the use of product standards that restrict the application of specific high emitters, or through architects and engineers specifying the use of low-emitting building materials. Levin (1989b) has developed sample specifications to avoid some pollutant sources. Reports suggest that catalogs will soon be published as a compendium of EPA and DOE research in this area to guide in the selection of materials with low pollution potential (McNall 1988). An example of a product standard is the HUD formaldehyde standard, which was discussed in Section 2.2.5.

Source Control - If sources of VOCs must be used, they may be controlled with spot ventilation, such as exhaust hoods over work benches, or with barriers. Effective barriers that block formaldehyde emissions include polyethylene plastic, vinyl linoleum, paints, shellac, varnishes, or lacquer. However, the coatings must be continuous and remain intact to be effective, and in their own right may be a source of VOCs. Barriers for other types of pollutants are not well understood.

Another approach to controlling occupant exposure to emissions is to minimize the quantity of the source introduced to the building and introduce the source in off-hours when the fewest people will be affected. One technique for reducing emissions is to condition building materials or furnishings before installation. Conditioning could be accomplished by airing out material, possibly in combination with elevated temperatures (Seifert et al. 1989).

Pollutant Removal - Ventilation is a good method of dissipating pollutants. Sheldon et al. (1988a) found that the half-lives of the compounds found in two new office buildings ranged from 2 to 8 weeks, and the time for the building to approach outdoor VOC concentrations would range from 3 to 12 months. Sheldon states that these findings compare well with the Scandinavian decision to require 100% makeup air for the first 6 months of a new building's life (Sheldon et al. 1988a).

Girman et al. (1987, 1989) report on a process called "bake-out" as a potential mitigation technique for high VOC levels in new commercial buildings. The procedure involves subjecting the indoor environment to high temperatures and high ventilation rates to encourage the emission and removal of VOCs. A building in San Francisco was heated to a maximum of 39°C (102°F) for 24 hours, at a ventilation rate of 1.59 air changes per hour (ACH). Fourteen VOCs were quantified. The total VOC concentration increased by 400% during the bake-out, and was reduced to 71% afterwards. Girman concludes that longer bake-out periods may be more effective. In addition to demonstrating the effectiveness of this mitigation technique, Girman's work underscores the important role of temperature, aging, and ventilation in influencing VOC concentrations.

2.3 RESPIRABLE SUSPENDED PARTICULATES

Respirable suspended particulates (RSP) are particles or fibers in the air that are small enough to be inhaled. They are a broad class of chemically and physically diverse substances and can be in either a solid or liquid phase, or in combination. They are generally less than 10 μm in diameter (DOE 1986).

Respirable suspended particulates are generated from building materials (fiberglass fibers, cellulose fibers, asbestos fibers), combustion devices (gas appliances, gas hot water heaters, and boilers), occupant activities (tobacco smoke, resuspended dust), and infiltration from outdoor sources (atmospheric dust, combustion emissions from mobile and stationary sources). However, the largest single source of RSP in the indoor environment is tobacco smoke (Mueller Associates, Inc., et al. 1987; Turk et al. 1986).

Asbestos is a collective term for a variety of asbestiform minerals, each of which satisfies a particular industrial-commercial need. These include chrysotile, anthophyllite, riebeckite, cummingtonite-grunerite, and actinolite-tremolite. Commercial names for the first four of these are chrysotile, crocidolite, anthophyllite, and amosite. These minerals differ in chemical composition, but all are silicates. Chrysotile accounts for over 95% of the asbestos sold in the United States (Godish 1989). Asbestos fibers are characterized by their small diameter, high length-to-width ratio, and great strength and flexibility.

Health risks from RSPs can be attributed to either their intrinsic toxic chemical or physical characteristics, as in the case of asbestos, or to the particles acting as a carrier of adsorbed toxic substances, as in the case of radon (see Section 2.6). Because they can deliver a high concentration of harmful substances deep in the lungs, RSPs present a health risk out of proportion to their concentration in air. Once in the lungs, they have a long residence time, affording a greater opportunity for surrounding tissue to become affected.

2.3.1 Pollutant Sources and Source Strength

Indoor sources of RSPs include combustion sources (fireplaces, wood stoves, unvented gas appliances), asbestos construction materials, and house dust. Other incidental indoor sources include the use of aerosol sprays, and wear and sloughing of building material. In the office environment, photocopy dust may also contribute to elevated levels of RSPs (Turk et al. 1986; Mueller Associates, Inc., et al. 1987). Some RSPs are from outdoor sources and enter the building interior via the HVAC system or through natural ventilation and infiltration.

Combustion Byproducts - Wood stoves and gas burners contribute to indoor RSP levels. Gas ovens and gas and kerosene space heaters contribute very little to overall RSP levels. Table 2.11 presents a summary of the particulate emissions rates for various combustion sources.

Tobacco Smoke - More than 95% of the particulates produced from tobacco smoking are from sidestream smoke. Sidestream smoke arises from smoldering tobacco. Particulate emissions from sidestream smoke are estimated to be as much as 30 to 40 mg per cigarette (Mueller Associates, Inc., et al. 1987).

TABLE 2.11. Pollutant Emission Rates (adapted from Mueller Associates, Inc., et al. 1985)

Source	Appliance Type	Emission Rate (mg/hr)
Kerosene Space Heaters	Radiant Convection	0.13 - 0.16 <0.03 - 0.034
Gas Space Heaters	--	0.21 - 3.23
Wood Heater		2.6
Gas Appliance Oven	Range (1 burner)	1.9 - 30 118 - 0.126

Fiberglass and Cellulose Fibers - Fiberglass RSP can be generated from the insulation used in ventilation ducting. Gamboa (1988) found that fiberglass fibers released from Type 475 duct board and fiberglass duct liner were typically well below the 3.0 fibers/cm^3 permissible exposure limit proposed in 1977 by NIOSH. However, these releases still resulted in fiber levels approximately twice normal background concentrations.

Asbestos - Commercial use of asbestos is extensive, with over 3000 applications, including fireproofing, thermal and acoustical insulation, acoustical plaster, friction products such as motor vehicle brake shoes, reinforcing material in cement water pipe, and roofing and floor products. Asbestos can be found in most large public access buildings, including schools, office buildings, and commercial buildings. Vinyl floor covering may also contain asbestos fibers, but historically it is not considered a significant source of asbestos because of the bound nature of the material (Godish 1989), except when it is disturbed, as during remodeling or removal.

Asbestos is used as thermal insulation around steam lines and boilers. In this situation, it is less of a threat for fiber release into occupied spaces and exposure to humans, because it is usually covered with a rigid cloth or paper material and is often inaccessible to most building occupants.

Asbestos release to the indoor environment depends on the cohesiveness of the asbestos-containing material (ACM) and the intensity of the disturbing force. Most contamination is episodic and activity-related (Mueller Associates, Inc., et al. 1987). The greatest threat for human exposure is from fiber release from surface materials as they age and the binding material weakens.

Prior to 1973, it was a common practice to apply ACMs by spraying or troweling to a variety of building surfaces, notably ceilings and steel beams. The use of such friable ACMs as a fire/heat retardant in public schools and other large buildings has been banned by the EPA in an effort to reduce community exposure to asbestos fibers during building demolition (Godish 1989).

2.3.2 Concentrations in Commercial Buildings

Generally, RSP concentrations are greater indoors than outdoors. Typical indoor RSP concentrations range from 100 to 500 $\mu g/m^3$, with the highest concentrations occurring in areas with tobacco smoking. Although wood and kerosene-burning appliances can contribute to higher levels of RSP in the indoor environment, other sources such as cooking, vacuum cleaning, aerosol spray products, and other activities are significant sources of indoor RSP (Saxton 1984).

Combustion Byproducts - A study conducted in 10 paired homes in Boise, Idaho, found that normally functioning wood stoves contributed 4 to 16 $\mu g/m^3$ to total RSP levels. Damaged or improperly operated stoves contributed over 70 $\mu g/m^3$ to the RSP levels (Highsmith et al. 1988). The Mueller Associates, Inc., et al. (1987) reported RSP levels in residences using wood-burning appliances ranging from 10 to 100 $\mu g/m3$.

Although indoor concentrations of RSP in wood-burning residences tends to be higher than in non-wood-burning residences, the difference is not statistically significant (reflecting the high variability in indoor concentrations).

Kerosene heaters can raise short-term concentrations to 100 to 300 $\mu g/m^3$ (Godish 1989), depending on the type, age, and use pattern.

Tobacco Smoke - In a study conducted in 38 commercial buildings in the Pacific Northwest, the geometric mean of RSPs in all buildings was 24 $\mu g/m^3$. The geometric means for smoking areas and nonsmoking areas were 44 $\mu g/m^3$ and 15 $\mu g/m^3$, respectively. Fourteen of the 70 smoking sites and only one of the 106 nonsmoking sites exceeded the ASHRAE annual exposure guideline of 75 $\mu g/m^3$. The highest level of 308 $\mu g/m^3$ occurred in a Portland office building cafeteria where smoking was allowed (Turk et al. 1986).

Even in buildings with local high indoor RSP levels due to tobacco smoke, nonsmoking areas of the same building can be quite low. In the above study, in only one building were RSP concentrations in the nonsmoking area within a factor of 2 of the smoking area concentrations. Local exhaust, ventilation near smoking areas, dilution by large building volumes, and removal mechanisms significantly reduce RSP concentrations in areas away from the source (Turk et al. 1986). However, in buildings with smaller volumes and poor ventilation, isolating smoking areas may have little effect at limiting the distribution of RSPs.

Fiberglass and Cellulose Fibers - No information could be located that specifically addressed concentrations of fiberglass and cellulose fibers.

Asbestos - Ambient concentrations of asbestos range from 0.00005 to 0.00022 fibers/cm^3. Concentration in public buildings and schools ranged from 0.00026 to 0.0040 fibers/cm^3 (Godish 1989). In nonresidential situations, asbestos concentrations range from 0.02 fibers/cm^3 (falling from exposed asbestos ceiling) to 17.7 fibers/cm^3 (during ceiling repair). During the mixing of drywall taping compounds, fiber counts measured were found to be from 7 to 12 times greater than the current occupational standard (Mueller Associates, Inc., et al. 1987).

2.3.3 Environmental Factors

Natural removal mechanisms for RSP include physical deposition, chemical transformation, and particle agglomeration. Chemical transformation of the nonfibrous particulates increase with increasing temperature. Higher relative humidities encourage the agglomeration of some airborne RSP into larger particles, accelerating their settling rate, as well as converting them into particles that are less likely to be inhaled. Although the larger particles may be inhaled, they will be deposited in the nose/mouth and not deep in the lungs.

2.3.4 Standards, Guidelines, and Regulatory Effects

Except for asbestos, a specific U.S. standard for RSP does not exist. The Occupational Safety and Health Administration (OSHA) has set an 8-hour time-weighted average limit of 5000 $\mu g/m^3$ for respirable inert or nuisance dust. The EPA established a 24-hour ambient air quality standard for particulates less than 10 μm in diameter (referred to as PM_{10}) at 150 μ/m^3. The ASHRAE total suspended particulate (TSP) annual exposure limit is 75 $\mu g/m^3$. Because RSP is a subset of the TSP, the ASHRAE limit can also be used as a conservative guideline for RSP.

The OSHA-established asbestos standard is less than 2 fibers longer than 5 μm per cm^3 for an 8-hour time-weighted average.

2.3.5   Interactions with Energy Conservation Measures

Building/system components that most significantly influence the indoor levels of pollutants include 1) envelope construction material and technique, 2) infiltration and natural ventilation, and 3) control and operation of the mechanical ventilation system. Energy conservation measures (ECMs) that alter or in any way affect any of these components will, either directly or indirectly, affect indoor RSP levels. A direct effect is one in which the ECM either increases or decreases the source emanation rates. An indirect effect is one in which the pollutant concentrations are increased or decreased. For example, in the absence of a strong indoor source (e.g., tobacco smoke), any ECM that results in a net reduction of natural infiltration will likely reduce indoor RSP concentrations due to the reduction of the rate of infiltration of outdoor particulates (i.e., fugitive dust) to the indoor environment. On the other hand, when major indoor sources are present, ECMs that reduce the rate of infiltration will most likely result in greater indoor RSP concentrations because of reduced dilution with fresh air.

In existing, older commercial buildings, any activity that disturbs existing ACMs will result in increased levels of asbestos fibers. Although the disturbing activity is only temporary, resuspension of the fibers due to normal occupant activity can result in elevated levels of RSP.

The paragraphs below describe the effects of ECMs on indoor levels of RSP by building system component.

Building Envelope - Installation of additional insulation in the walls, ceiling, roof, foundation, or slab can directly increase the levels of RSP/fibers by increasing the source of fiberglass/cellulose material in the building.

Any ECMs designed to decrease infiltration/exfiltration by modifying the building envelope will indirectly affect the levels of RSP. For RSPs with indoor (outdoor) sources (such as combustion byproducts), the RSP levels will be increased (decreased) due to the decreased rate of dilution of the pollutant with fresh air. These ECMs include 1) weatherstripping and caulking, 2) installation of storm doors and windows, 3) enclosing loading docks with shelter and seals, 4) installation of vestibules, and 5) sealing vertical shafts.

HVAC System - The only HVAC ECM that will affect indoor RSP/fiber levels will be the installation of duct insulation. One study suggests that the relative increase in fiber levels can be large, although normal background levels are generally low and it is unlikely any standards will be exceeded.

Ventilation - Recirculation of exhaust air using activated carbon filters has the potential for directly increasing RSP/fiber levels by increasing the opportunity for fiber release within the building. This same ECM, however, will also remove RSPs (including fibers) from the indoor air as they are recirculated through the ventilation system.

The installation of vortex hoods in the kitchen areas of restaurants will effectively remove much of the combustion byproduct RSPs generated by the gas burners, thereby reducing RSP levels in the remainder of the building.

2.3.6 Specific Mitigation Techniques

Mitigation techniques for controlling RSPs fall into three broad categories: source control, ventilation, and air cleaning. A generic description of these techniques is presented in Section 4.

More specific control technologies for RSP include the following (Mueller Associates, Inc., et al. 1987):

- Limit or eliminate smoking tobacco indoors.

- Make sure wood stove doors and flues do not leak.

- Supply outdoor air directly to wood stove and fireplace firebox.

- Use electrostatic precipitators and high-efficiency particulate air (HEPA) filters as effective removal media.

- Change air filters regularly.

- Vent combustion appliances outdoors.

Air cleaners for particulate control are available as both in-duct devices or stand-alone devices for single room use. Particulate air cleaners can be divided into two groups: mechanical filters and electrostatic filters. Mechanical filtration is generally accomplished by passing the air through a fibrous medium. Electrostatic filtration operates on the principle of attraction between opposite electrical charges. Ion generators and electrostatic precipitators use this principle for removing particles from the air.

For asbestos, source control is the only acceptable mitigation option. This is a reflection of its hazardous nature and the compelling need to prevent asbestos fibers from becoming airborne. Special asbestos abatement measures include repair, enclosure, encapsulation, and removal. The EPA considers removal the only permanent solution and recommends the use of enclosure or encapsulation only under limited circumstances (Godish 1989).

2.4 BIOLOGICAL CONTAMINANTS

Biological contaminants are either living organisms (bacteria, fungi, viruses, amoebae, algae, and pollen grains) or the byproduct of living organisms. The organism byproducts include plant parts; insect parts and wastes; animal saliva, urine, and dander and human dander; and a variety of organic dusts (Godish 1989).

The source of biological contaminants can be either in the indoor environment or outdoors, and transported indoors via infiltration and/or through the building's ventilation, air-conditioning, and humidifying systems. Indoor surface contamination by bacteria, fungi, insects, arachnids, or other biological particles usually will not have adverse health effects until the particles become airborne and are inhaled (Burge 1985).

Biological contaminants can produce direct toxicity (as in the case of some molds), or they may be pathogenic (provoking infectious disease) or allergenic (provoking hypersensitive or allergic disease). The majority of biological contaminants are nonpathogenic and cause adverse health effects only in individuals who are sensitive to that particular contaminant, such as allergic reaction to plant pollen. Even many pathogenic microorganisms are able to infect only susceptible individuals (Burge 1985). Even though there is a lack of understanding about the magnitude of the health effects caused by exposure to biological contaminants, the consequences of some are appreciable. For example, the most common respiratory disease attributable to allergens is allergic rhinitis (allergies), which affects approximately 15% of the U.S. population (Mueller Associates, Inc., et al. 1987).

Hospitals and other exceptional "clean" room environments have their own unique requirements and are not addressed in this document.

2.4.1 Pollutant Sources and Source Strength

The paragraphs below describe the various sources of biogenic contaminants. No information is available on the source emanation rates.

The two most important allergen contaminants found in indoor air that are known to cause both allergies and asthma are produced by dust mites and fungi (Godish 1989). The most severe indoor biological contamination results from growth of the offending organism on surfaces within structures, such as cooling towers and condensate drip pans on air conditioners. Virtually any surface that includes both a carbon source and a supply of water will support the growth of some microorganism.

Many buildings with biological contamination can trace the probable cause to the lack of proper maintenance of the HVAC system. Typical maintenance problems include condensate drains and drip pans that are not cleaned, filters that are not cleaned or replaced, and dirt and biological growth in the duct system (Robertson 1989).

The primary source of indoor bacteria is the human body. Although the major source is the respiratory tract, it has been shown that 7 million skin scales are shed per minute per person, with an average of four viable bacteria per scale (Mueller Associates, Inc., et al. 1987).

Cool mist vaporizers and nebulizers can produce heavily contaminated aerosols. Other potential sources of indoor bacterial aerosols include ice machines, commercial dishwashers, air conditioners, and humidifiers (Mueller Associates, Inc., et al. 1987). Flush toilets and urinals surely aerosolize bacteria, but do not pose great disease risk unless they are heavily

contaminated and poorly maintained (Burge 1985). Pathogenic microorganisms
are spread by the use of the same air-conditioning and humidifying equipment
that can be the source of the contaminant.

High mold levels in office buildings can result from mold growth in
cooling coil drip pans found in HVAC systems, in evaporative condensers, and
on water-damaged materials.

Many bacteria, including *Legionella pneumophila* (Legionnaires' Disease),
thrive in outdoor natural reservoirs, and infection is possible through
inhalation of contaminated outdoor air. The most common mechanism for spread-
ing bacteria is air-cooling equipment that is contaminated and produces
concentrated bacterial aerosols (Mueller Associates, Inc., et al. 1987).

Most fungi are primary decay organisms and are abundant on dead or dying
plant and animal materials. Thus, natural wildlife habitats, parks, and green
spaces near buildings will surely lead to some indoor contamination and
possible risk of infection. The extent of contamination from fungi is
unknown, but some are known to be highly infectious.

2.4.2 Concentrations in Commercial Buildings

Building entry of biological contaminants from the outside is through
cracks, windows and doors, or through ventilation, air conditioning, and
humidifying systems. Rate of entry is determined by indoor/outdoor pressure
differentials, quantity, and type of ventilation, wind conditions, temperature, and humidity (Baechler et al. 1989).

During the growing season, outdoor fungus spore levels exceed those
indoors. Only in cases of extreme indoor contamination do indoor spore levels
exceed outdoor levels (Burge 1985).

Extensive, quantifiable data on indoor concentrations in the case of
microorganisms, allergens, and pathogens is limited (Mueller Associates, Inc.,
et al. 1987). In one study, air sampling in seven office buildings in New
York City for mesophilic fungi, Morey and Jenkins (1989) found that in most
buildings the indoor concentrations of fungi during normal periods of
occupancy were approximately 10% to 25% of coincident outdoor concentrations.
However, during periods when furnishings and the HVAC system components were

disturbed, concentrations of airborne fungi in the immediate vicinity generally exceeded outdoor levels by a factor 1.5 to 13. In one building, the indoor concentrations exceeded outdoor levels by a factor of 100. Outdoor fungi levels ranged from a low of 36 colony-forming units per cubic meter of air (cfu/m^3) to over 800 cfu/m^3.

2.4.3   Environmental Factors

Ideal growing conditions for microorganisms such as spores, molds, and fungi include moisture, humid air, and warm temperatures. Microorganisms require humidity in the range of 25% to 70% to increase in number. Above 70% is optimal for growth of molds and bacteria and for the survival of airborne pathogens (Mueller Associates, Inc., et al. 1987; Burge 1985). The lowest relative humidity at which a fasting mite can sorb sufficient water to maintain water balance is 70% to 73%. This condition must be met or mite population cannot survive (Godish 1989).

Skov and Valbjorn (1987) and Valbjorn and Skov (1987) found no association between SBS and ventilation characteristics, but did find strong positive correlations with the age of the building, the total weight and potential allergenic portion of floor dust, the area of fleecy material, the open shelving per cubic meter of air, and the air temperature.

Air currents produced by convection from radiant heat and by air mechanically circulated by forced air systems are more than adequate to spread dust (including entrained biological particles), as well as mobilizing surface growth (Burge 1985).

2.4.4   Standards, Guidelines, and Regulatory Effects

There are no standards or guidelines established for biological contaminants in commercial buildings. However, for some hospital operating rooms, the upper concentration limit for particulates containing viable bacteria has been set at 175 particles/m^3 (Morey 1985).

2.4.5 Interactions with Energy Conservation Measures

Energy conservation measures that directly or indirectly affect the sources of biological contaminants or the concentrations of contaminants in the indoor air are described below by building/system component.

Building Envelope - Any of the building envelope ECMs that are designed to reduce infiltration/exfiltration through envelope sealing will likely have the net effect of reducing indoor levels of biological contaminants from outdoor sources by reducing the transport of contaminated outdoor air to the indoors.

For indoor sources, shell-tightening measures will likely increase indoor concentrations by reducing contaminant dilution. Research conducted on military barracks suggests that tight buildings with closed ventilation systems significantly increase the risks of respiratory-transmitted infections among congregated, susceptible occupants (Brundage et al. 1988).

Potential ECMs that will reduce infiltration include

- caulking and weatherstripping
- installing storm doors and windows
- enclosing loading docks with shelters and seals
- installing vestibules
- sealing vertical shafts.

HVAC System - Many of the potential HVAC system ECMs will directly increase the potential for biogenic contamination, particularly for the bacteria contaminants that develop in water reservoirs. These ECMs are

- replacing air-cooled condenser with wet cooling tower
- installing roof spray systems
- installing spot cooling.

These three ECMs either use water directly or will indirectly generate water as condensate that can collect in drip pans, basins, and other containers. If not properly maintained, the water reservoirs will become contaminated with microorganisms.

Ventilation - Installing evaporative cooling of outside air will directly increase the threat of bacterial contamination by increasing the breeding ground for bacteria. Any of the ECMs that will reduce ventilation and/or increase the use of untreated recirculating air could amplify the concentration of microorganisms. These include

- reducing ventilation automatically during unoccupied periods

- recirculating exhaust air using carbon activated filters

- installing dual-speed fans (if the net effect is an overall net reduction in air circulation).

Installing a carbon dioxide (CO_2)-controlled ventilation system may decrease levels of biogenic contaminants during periods of building occupancy by increasing ventilation rates over "normal" rates.

2.4.6 Specific Mitigation Techniques

Biological contaminants are best controlled by applying a variety of source control measures, primarily focusing on restricting the availability of moisture necessary for microorganism growth. Removal of nutrients needed for microbiological growth may also control contamination. Air cleaning can be of limited effectiveness (Burge 1985; Mueller Associates, Inc., et al. 1987; Baechler et al. 1989).

Source prevention is essential for control of microorganisms. Source control strategies fall into two broad categories: 1) initial system design that minimizes the potential for microbial growth and subsequent contamination of building air-handling systems, and 2) preventive maintenance, including adherence to routine inspection and proper maintenance of all existing water-using or -generating systems associated with the HVAC system.

From a systems perspective, humidification units based on recirculated water should not be used. Steam as a moisture source is preferred (Godish 1989). If cool water humidifiers are to be used, water should originate from a potable source and run to a drain instead of being recirculated. The system should be regularly inspected, cleaned, and disinfected.

Cool-mist vaporizers, such as found in grocery store produce areas, have recently been linked with the bacteria *Legionella pneumophila*.

Elimination of the use of water-spray systems in office building HVAC systems, restriction in the use of humidifiers, proper location of cooling towers with respect to building vents, and the utilization of available filtration-air disinfection devices have been advocated as control measures for specific indoor environments.

Preventive maintenance or housekeeping activities, together with protection from floods and moisture incursion, will reduce the chance of microbial contamination in the indoor environment. Stagnant water should not be allowed to accumulate in cooling coil condensate drip pans. Drip pans should be properly inclined to provide continuous drainage (Godish 1989). Plumbing and other fluid systems should be maintained so that no water incursions occur in the building. If incursions do occur, they should be rapidly cleaned up and water-damaged materials discarded. Water-damaged material (such as ceiling tiles and carpeting) that has become contaminated with microorganisms should be discarded and replaced.

Relative humidity less than 50% is necessary to keep mold levels below average (Burge 1985). Maintaining relative humidity levels as low as reasonably achievable, while consistent with occupant comfort, can be an effective control strategy. In kitchens and other normally high moisture areas where it is not possible to lower humidity levels, maintenance and cleaning of water reservoirs becomes even more important as a control strategy to prevent microorganism proliferation.

Mitigation measures involving actual air cleaning fall into two categories: reducing outdoor influx of contaminants and removing contaminants from indoor air. However, air cleaners produce limited results. In most situations where air cleaners are used, there are continuing sources (outside air, dust, people) and the cleaners reach a balance between the effective rate of removal and the rate of contaminant generation. Unfortunately, the balance is usually not on the side of clean air (Burge 1985).

Antimicrobial agents are less effective in preventing contamination than the basic environmental changes discussed above. They can be useful in cleaning known sources, but unless the basic conditions causing the initial contamination are changed, recontamination will occur (Burge 1985).

2.5 COMBUSTION GASES

Combustion gases, such as carbon monoxide (CO), carbon dioxide (CO_2), nitrous oxide (NO), nitrogen dioxide (NO_2), and sulfur dioxide (SO_2) can be introduced into the indoor environment by a variety of either indoor or outdoor sources. For the purposes of this report, environmental tobacco smoke (ETS) is included along with the other combustion byproducts that contribute to the gaseous contamination of indoor air. Smoking of tobacco products indoors is the major source of combustion-generated contaminants found in indoor air (Godish 1989). More than 2000 gaseous compounds have been identified in cigarette smoke (Mueller Associates, Inc., et al. 1985, 1987).

2.5.1 Pollutant Sources and Source Strength

Combustion gases are introduced into the indoor environment from either indoor combustion sources or from outdoor sources, such as attached parking garages. The description of the sources of combustion gases presented below is adapted from Mueller Associates, Inc., et al. (1985).

Nitrogen Oxides (NO_x) - Indoor levels of NO_x are primarily affected by gas stoves, unvented kerosene and gas space heaters, and, to a lesser degree, tobacco smoking. Wood stoves do not usually produce significant emissions of NO or NO_2. Emission rates for NO_x from appliances are shown in Table 2.12. The emission rate of NO_x from tobacco smoking is 0.06 to 0.73 mg/cigarette. Infiltration of NO_x from outdoor sources is significant in heavily populated, industrialized urban areas where fossil fuel combustion and vehicles emissions are major outdoor sources of NO_x.

Carbon Monoxide - Carbon monoxide is formed whenever fuel is incompletely burned. The CO concentration in the indoor environment is mostly a result of combustion appliances such as gas stoves, portable kerosene and gas space heaters, and wood stoves. Tobacco smoking is a lesser source. Outdoor sources of CO, including emissions from vehicles in attached garages, also contribute to indoor levels of CO. The CO emission rate for tobacco smoking is 52 to 105 mg/cigarette.

TABLE 2.12.   Emission Factors for Gas Appliances (adapted from
              Borrazzo et al. 1985)

Appliance	CO (μg/sec)	NO (μg/sec)	NO$_2$ (μg/sec)
Range Burner			
1-hour ave - high	350	33	45
- medium	390	25	38
- low	360	10	23
steady-state - high	260	37	43
- medium	360	26	38
- low	320	10	23
Oven			
400°F, 0-15 min	760	100	108
400°F, 15-60 min	84	36	26
500°F	86	58	40
Pilot lights	14	1	2
Furnace	<10	2	1
Water Heater	<10	38	6
Clothes Dryer			
Fugitive emissions	<10	2	1
Exhaust emissions	110	110	71

Carbon Dioxide - The single greatest contributor to indoor CO_2 is human metabolic activity. Significant quantities of CO_2 are also generated by gas stoves and portable kerosene and gas space heaters, and can be introduced to the indoor environment if not vented properly. Wood stoves and tobacco smoking produce smaller amounts.

Sulfur Dioxide - Indoor SO_2 is usually a result of infiltration of SO_2-contaminated outdoor air, primarily from stationary fuel combustion and industrial processes. Vehicle emissions contribute a smaller amount. The only significant potential indoor sources are portable kerosene space heaters and leaky flues in buildings where sulfur-containing fuel is burned. Emission rates for kerosene heaters range from 30 mg/hr to over 100 mg/hr.

Benzo[a]pyrene (BaP) - Indoor concentrations of BaP are a result of wood combustion and tobacco smoking. Outdoor concentrations of BaP as an indoor source appear to be insignificant.

The rates at which combustion gases are introduced into the indoor environment from appliances are summarized in Table 2.12.

2.5.2 Concentrations in Commercial Buildings

For intermittent sources, source emission rate, building volume, and rate of mixing are the primary factors affecting peak concentrations. For continuous sources, source emission rate and duration, the outdoor concentration, and air exchange rate are the primary factors affecting long-term average concentrations (Mueller Associates, Inc., et al. 1985). Numerous studies do not offer clear evidence that combustion gases can reach concentrations substantially higher than ambient concentrations (Spengler and Cohen 1985).

Nitrogen Oxides - In a Bonneville study of 38 commercial buildings in the Pacific Northwest, a total of 245 sample locations in 33 buildings were tested for NO_2. The geometric mean of all sample locations was 18.3 ppb, with a standard deviation of 1.7 (9% of the mean). Smoking areas had a geometric mean of 20 ppb. Only two monitoring locations in two different buildings had NO_2 levels greater than the EPA ambient air quality standard of 50 ppb. One was a 58-ppb average with a building average of 22 ppb, and the other was in a building known to be contaminated with vehicle exhaust emissions (Turk et al. 1986).

Carbon Monoxide - In the same Bonneville study, only six of 32 buildings monitored for CO had a time weighted average concentration greater than the minimum detectable level of 2 ppm (Turk et al. 1986). For these six buildings, the means ranged from 2.1 ppm to 3.3 ppm. Three of the six buildings had underground parking.

Turner (1988) seldom found CO levels greater than 9 ppm in buildings monitored in Australia, the United Kingdom, and the United States.

In a survey of CO concentrations in 25 individual stores in a large shopping center in Honolulu, Hawaii, OSHA's 1-hour standard of 200 ppm was not exceeded at any of the sample outlets on any survey date. The National

Ambient Air Quality Standard (NAAQS) of 35 ppm for 1 hour was exceeded on at least 18 of the 30 (60%) survey dates. Hawaii's Ambient Air Quality Standard (AAQS) of 9 ppm was exceeded on every survey date at many outlets. The OHSA's 8-hour CO standard of 50 ppm was not exceeded on any survey date. However, 21 of the 25 outlets (84%) had 30-day averages exceeding the NAAQS of 9 ppm, and 23 of 25 outlets (92%) had 30-day CO averages exceeding Hawaii's AAQS of 4.5 ppm. Average CO levels were 3 to 23 ppm inside the 25 street-level outlets. Motor vehicle emissions of CO were believed to be the primary source. In restaurant and fast-food establishments, gas-fired stoves and ovens were expected to be another source of CO (Flachsbart and Brown 1985).

Carbon Dioxide - In a study of commercial buildings in the Pacific Northwest, 39 8-hour average CO_2 measurements were made in 37 buildings. Concentrations were found to be dependent on occupancy levels. Only one 15-minute CO_2 reading in a crowded elementary school classroom exceeded 1000 ppm (1290 ppm). Periods of low occupancy (recess and lunch) were evident in the time series data. Four buildings had instantaneous CO_2 maxima over 800 ppm. Eight-hour averages ranged from 337 ppm to 840 ppm (Turk et al. 1986).

General office area CO_2 levels have been found by Turner (1988) to range from 350 ppm to 2000 ppm.

Sulfur Dioxide - No information was found describing SO_2 concentrations in commercial buildings, but they are believed to be generally low. The only exception would be if kerosene space heaters are used extensively.

2.5.3 Environmental Factors

Two major environmental factors that can influence the contribution of ambient (outdoor) levels of combustion-related gases to indoor contaminant levels and location and time (Mueller Associates, Inc., et al. 1985):

- The location of a building relative to major outdoor sources can affect indoor air quality. Depending on the emission rates, air pollutants from stacks, flues, vents, and cooling towers can also affect the indoor air quality. Buildings located near major streets or highways may be affected by the gaseous pollutants generated from vehicles. In general, urban and industrialized areas have higher concentrations than rural areas.

- Temporal fluctuations in concentration of outdoor gaseous pollutants will also influence indoor concentrations. Diurnal or seasonal patterns and day-to-day variations in the weather and daily
  variations in emission rates all determine ambient concentrations.

Pollutant reactivity can be, in some cases, as important as air exchange
rates in reducing indoor combustion gas concentrations, especially when the
air exchange rate is low. Nitrogen dioxide is removed through chemical
reaction. Carbon monoxide may be adsorbed on indoor surfaces, and particulate
matter will be deposited on indoor surfaces. The rate at which reactions
occur varies for each pollutant and is influenced by temperature, moisture,
presence of other compounds, and types of surfaces found (Mueller Associates,
Inc., et al. 1985).

Combustion byproducts may be removed from the indoor atmosphere by a
number of mechanisms other than air exchange. Individual pollutants may be
reduced by one or more of the following physical or chemical reactions:
adsorption, absorption, conversion, and deposition. Indoor-combustion-
generated pollutants are modified by these reactions as follows:

- nitrogen oxides - NO_2 is removed through reaction; NO may be
  adsorbed on surfaces.

- carbon monoxide - CO is comparatively inert, but may be adsorbed on
  indoor surfaces.

- sulfur dioxide - The SO_2 concentration may be reduced by the
  oxidation of SO_2 to sulfate particles or by dry deposition on
  surfaces. SO_2 interacts with airborne particles primarily by
  adsorption.

A number of environmental factors will affect the rate at which
reactions occur indoors, which will, in turn, affect the rate of removal of
pollutant concentrations.

- temperature - The rates of most chemical reactions increase with
  temperature. The rate of reaction will roughly double for each
  10°C rise in temperature

- moisture - The moisture in the air can enhance pollutant deposition
  rates. This is true for gases such as SO_2.

- solar radiation - Some reactions require the presence of sunlight.
  Decomposition of NO_2 to NO and ozone (O_3) requires sunlight.

- presence of other compounds - Reactions are affected by the presence or absence of certain compounds.

- surfaces - Some air pollutants differ in stability between indoors and outdoors because of the variety of surfaces found indoors on which reactive compounds adsorb and decompose. For example, O_3 is decomposed especially quickly indoors. This is also somewhat true for SO_2 and NO_2, while CO and NO are less reactive.

2.5.4 Standards, Guidelines, and Regulatory Effects

The Occupational Safety and Health Administration has established regulations governing the exposure of workers to toxic and hazardous substances in the workplace. These standards are listed in Table 2.13.

2.5.5 Interactions with Energy Conservation Measures

In public buildings without indoor combustion sources, no significant increase in indoor-combustion-related pollution levels is expected to occur with reduced ventilation rate. A single exception is CO_2 from human metabolic activity, which was still below levels considered a health hazard. Indoor air quality improved for outdoor generated sources.

In general, lower air exchange rates result in higher concentrations of contaminants from indoor sources. However, concentrations of reactive pollutants such as NO_2 and RSP may not be directly related to the air exchange rate. The air change rate plays an increasing role with time and has relatively little effect on initial peaks in the indoor pollutant concentrations (Mueller Associates, Inc., et al. 1985).

TABLE 2.13. OSHA Standards for Combustion Gases

Pollutant	Standard ppm	Standard mg/m^3	Time Interval
CO_2	5000	9000	8-hour time-weighted average
CO	50	55	8-hour time-weighted average
CO	200	220	1-hour average
NO	25	30	8-hour time-weighted average
NO_2	5	9	Ceiling value
SO_2	5	13	8-hour time-weighted average

Specific energy conservation measures that will affect combustion-related contaminant levels in commercial buildings are described below.

Building Envelope - Any ECMs that will tighten the building envelope will reduce the contribution of outdoor contaminants to the general level of pollutants indoors. However, in the presence of strong indoor sources, these same ECMs will tend to increase indoor contaminant levels due to reduced exfiltration rates and dilution with fresh air. These ECMs include

- caulking and weatherstripping
- installation of storm doors and windows
- enclosing loading docks with shelters and seals
- installing vestibules
- sealing vertical shafts.

HVAC System - None of the HVAC system ECMs will affect indoor concentrations of combustion gases.

Ventilation - The installation of variable rate, CO-controlled exhaust ventilation systems in covered parking garages will reduce indoor levels of CO by reducing the primary outdoor source of CO in buildings with attached parking garages. Installation of vortex hoods in the kitchen areas of restaurants will reduce levels of contaminants from gas stoves by effectively removing the contaminants at the source. Other ECMs that affect the ventilation system will have only minor impacts on indoor air quality.

2.5.6 Mitigation Techniques

In general, source removal is the most effective way to control indoor pollutant levels (Mueller Associates, Inc., et al. 1985). Increasing whole-building ventilation causes a smaller decrease in the concentration of a reactive pollutant such as NO_2 than nonreactive pollutants (Mueller Associates, Inc., et al. 1987). Local exhaust ventilation in the vicinity of a pollutant source is an effective source specific ventilation control technique (Kandarjian 1988). Air cleaning devices and the use of new, less polluting technologies can also be used to mitigate (Mueller Associates, Inc., et al. 1985).

It appears that the most practical long-term solution to minimizing the effects of tobacco smoking and to minimize the buildup of CO_2 from occupant metabolism is to ventilate office buildings to the ASHRAE Standard 62-1989 requirement of 20 cfm of outside air per person.

2.6 RADON

Radon is an inert, radioactive gas that occurs naturally in the environment as a decay product of radium. It, in turn, decays to form radioactive progeny that may attach to dust particles or remain unattached. If these progeny are inhaled, they can be drawn into the lungs, where they emit alpha energy which may lead to lung cancer.

Ambient concentrations of radon are generally low due to dilution with large volumes of outdoor air. Typical indoor concentrations can be as much as an order of magnitude greater than outdoors, and can exceed three orders of magnitude.

2.6.1 Pollutant Sources and Source Strength

Primary sources of radon and radon progeny in buildings are soil gas from the soil adjacent to the building, potable water, and building materials.

Soil - The amount of radon entering a building is dependent on the adjacent soil radium content and porosity, as well as the building construction type. Radon-bearing soil gas is transported into the building by pressure-induced convective flows. The EPA has estimated that the average soil contains about 1 ppm of uranium. Phosphate rock contains 50 to 125 ppm. Granite contains about 10 to 50 ppm in the northeastern United States and as much as 500 ppm in the western United States (Mueller Associates, Inc., et al. 1987).

The relative contribution of radon from soil gas to the total radon concentration in a building is proportional to the ratio of foundation area to building volume. The larger the ratio, the more significant the contribution of radon from soil gas. In residences, over 90% of all radon is from the soil. For large, multistory commercial buildings, soil gas is expected to contribute little to the overall radon levels (Godish 1989).

Water - Radon from ground water is found in certain potable water supplies, particularly in wells supplying small public systems where the

storage period is too short to allow for radon decay. Typical radon levels in
potable water fall below 2,000 picocuries per liter (pCi/l), but concentrations exceeding 100,000 pCi/l have been found in Maine (Mueller Associates,
Inc., et al. 1987). For typical water supplies, 10,000 pCi/l of radon would
contribute 1 pCi/l to the indoor air in residences (Mueller Associates, Inc.,
et al. 1987). In the Pacific Northwest, potable water is considered to be a
relatively minor source of radon.

Building Materials - Building materials are a major source of elevated
radon concentrations only if their radium content and emanation ratios are
very high. Ordinary building materials may be the dominant source of indoor
radon in buildings with low to moderate concentrations. This would be
especially true in multistory buildings in which the underlying soil contributes relatively little radon per unit volume. Concrete is the strongest radon
source; of the components in concrete, sand is the strongest source. Wood is
the lowest source of radon (Mueller Associates, Inc., et al. 1987).

2.6.2 Concentrations in Commercial Buildings

Although radon's behavior in commercial and multifamily buildings is
poorly understood, the results of a study conducted by Bonneville in the
Pacific Northwest showed that radon concentrations are low in a non-
representative sample of commercial buildings. Measured radon levels in 39 of
the 40 buildings studied were generally low; the geometric mean for all
buildings was 0.8 pCi/l. Only 6% of the sample was above the Bonneville
mitigation level of 5 pCi/l. The highest building average was 7.4 pCi/l in a
Cheney, Washington, office building. The Cheney building has a basement that
is partially exposed to bare soil and underground service tunnels with rock
block walls and floors. The main air handler fan for the building is located
in the basement and draws "fresh" air from the basement and tunnels (Turk
et al. 1986).

By comparison, in a study of 250 residences in the Northwest, the
average first-floor radon level was 1.2 pCi/l. Only about 4% of the homes
experience radon levels above the Bonneville action level guideline of 5 pCi/l
(Bonneville 1984).

2.6.3 Environmental Factors

The most significant environmental factors affecting radon concentration in commercial buildings are the weather and soil moisture. The primary mechanism for radon to enter the structure from the soil is through pressure-driven flow (Mueller Associates, Inc., et al. 1987). This flow is the result of a temperature-induced pressure difference across the building envelope and a top-to-bottom pressure difference induced by the wind. These pressure differentials result in a net upward movement of air in the building, with part of the replacement air coming from the building basement/foundation area.

Temperature, pressure, and moisture have been found to affect the rate of radon emanation from building materials. Of these three, moisture content appears to have the most significant effect on increased radon emanation rates (Mueller Associates, Inc., et al. 1987).

2.6.4 Standards, Guidelines, and Regulatory Effects

A number of organizations have recommended guidelines and standards for radon in residences and commercial buildings. Those relevant to commercial buildings are listed in Table 2.14.

TABLE 2.14. Radon Action Levels and Guidelines

Organization	Standard	Comment
National Council on Radiation, Protection and Measurement	8 pCi/l	Recommended action level for general population
ASHRAE and World Health Organization	5.4 pCi/l	Recommended exposure level in commercial buildings and residences
Bonneville	5 pCi/l	Action level for residential weatherization program
EPA	4 pCi/l	Guideline for indoor radon

Source: DOE 1986

2.6.5 Interactions with Energy Conservation Measures

Energy conservation measures that will directly or indirectly affect the indoor concentrations of radon in commercial buildings are described below.

Building Envelope - The impacts of building envelope ECMs on indoor levels of radon are not well understood. Measures that will decrease the infiltration of outdoor air may increase radon concentrations by reducing the rate of fresh-air dilution. However, those same measures may decrease indoor concentrations by blocking the radon entry points and reducing the pressure differences that drive radon entry. Specific ECMs are

- caulking and weatherstripping
- installation of storm doors and windows
- enclosing loading docks with shelters and seals
- installing vestibules
- sealing vertical shafts.

HVAC System - None of the HVAC system ECMs is expected to increase radon concentration. However, if the ECM should result in a lowering of the interior pressure and an increase in the pressure-driven flow into the building, radon entry will increase.

Ventilation - Ventilation system ECMs can result in either an increase or decrease in radon concentration, depending on the measure:

- Automatic reduction of ventilation rates during unoccupied periods will increase radon levels by decreasing the rate of dilution.

- Installing CO_2-controlled ventilation systems will result in decreased levels of radon by increasing the rate of dilution with fresh air.

- Recirculating exhaust air using activated carbon filters that are properly maintained will have a net reducing effect by removal of radon-carrying particulates.

- Installing attic ventilation will increase the transport of radon into the building interior by increasing the top-to-bottom pressure differential.

2.6.6 Specific Mitigation Techniques

The most effective mitigation technique for radon is source control. Incorporation of radon-resistant construction features into building design

before construction will be more cost-effective than remedial actions. Basic
preventive measures in new construction include the following (Mueller
Associates, Inc., et al. 1987):

- avoidance of high-radium content substances as components of
  building materials

- attention to site characteristics, such as local geology and soil
  permeability, which could indicate a potential for high levels of
  radon-bearing soil gases

- avoidance of sites containing waste materials that are high in radium

- construction techniques with methods that will minimize any high-
  permeability pathways between the soil and indoors including sealing of
  radon entry routes

- soil ventilation to draw soil gas away from the building

- crawl space ventilation to exhaust radon before it enters the occupied
  space of the building

- basement overpressurization to reverse normal pressure gradient to
  inhibit transport of soil gas into the basement

- air exchange rate increases

- air cleaning using HEPA or electronic filters to remove airborne
  particulates to which radon products attach.

3. Ventilation

This section presents a general overview of infiltration and mechanical ventilation in commercial buildings and summarizes the results of whole-building air exchange rate measurements.

3.1 CHARACTERIZATION OF VENTILATION

The overall exchange of air in buildings with fresh outside air is the result of the combination of infiltration, natural ventilation, and mechanical ventilation. Even in new buildings, infiltration can be a significant contributor to the total building air exchange rates. This is particularly true in the winter when the driving forces for infiltration are the greatest and the mechanical ventilation systems are operated with a minimum of outside air.

The processes of infiltration, natural ventilation, and mechanical ventilation are described in Sections 3.1.1 and 3.1.2.

3.1.1 Infiltration and Natural Ventilation

Infiltration is the flow of air through cracks, interstices, and unintentional openings in the building envelope. Natural ventilation is under the manual control of building occupants. It occurs through operable windows, doors, skylights, roof ventilators, stacks, and other planned inlet and outlet openings. It can be classified as "controlled" infiltration/exfiltration. The same forces that produce infiltration also drive natural ventilation.

Natural ventilation is more likely to occur during periods of moderate to warm outdoor weather conditions. During these periods, infiltration is at a minimum (low ΔT) and building occupants may open windows and other intentional openings to reduce indoor temperatures or create air motion to increase occupant thermal comfort (Godish 1989). Fortuitously, indoor levels of contaminants may also be reduced by occupants controlling natural ventilation to reduce that "stuffy" feeling of indoor air.

The quantity of air flowing through openings (either unintentional or planned) depends on both the dynamic pressure of the wind and buoyancy forces

54

resulting from indoor/outdoor temperature differences. The interaction of these forces is summarized below. It is beyond the scope of this report to provide an in-depth discussion of the interaction of these two forces. An excellent treatment of this subject is presented in Godish (1989) and Mueller Associates, Inc., et al. (1987).

The wind effect is due to airflow around a building creating regions in which the static pressure is either greater than (positive) or less than (negative) the pressure inside the building. Simplistically, wind pressures are positive on the windward side of the building and negative on the leeward side. Negative pressures also exist across the top of the building. Positive pressure differentials result in infiltration, and negative pressure differentials result in exfiltration.

The buoyancy force, or stack effect, is a result of the temperature differences between the inside and outside of the building. During the heating season, the warmer inside air is less dense and rises and exits the building near its top. It is replaced by cooler/colder outside air, which flows in near the bottom of the building. The stack effect increases with increasing building height and can be exceptional in tall buildings with vertical passages such as elevator shafts and stairwells.

The degree of air exchange due to infiltration or natural ventilation is dependent on the magnitude of the two driving forces. Infiltration is the greatest during the colder months of the year when indoor/outdoor temperatures are the greatest. During warm weather periods, when natural ventilation is most likely used, the indoor/outdoor temperatures are relatively small, and the stack effect is minimal, the wind effect is the dominant infiltration mechanism. Buildings that rely on natural ventilation to supply their ventilation air are at risk during periods of stagnant conditions.

3.1.2 Mechanical Ventilation

In the past decade, most large commercial buildings have been designed to provide year-round climate control. Because windows are usually sealed, occupant-controlled natural ventilation is restricted. Fresh, outdoor air is available only through a mechanical ventilation system.

Simply defined, mechanical ventilation is the forced movement of air by fans into and out of a building. The primary purpose of the mechanical ventilation system is to provide a healthy and comfortable indoor environment for building occupants. Other purposes include temperature and humidity control, improved thermal comfort, air exchange control, and exhausting smoke, waste heat, and toxic pollutants. Mechanical ventilation systems are designed to provide a range of ventilation rates. Most HVAC systems are designed to operate on a minimum of outside air--averaging 15% to 20% of total airflow. Through intentional control of the dampers, or even failure of the damper controls, the percentage of outside air introduced to the ventilation system can range from 0% to 100% (Godish 1989).

Two of the most commonly used HVAC systems are the constant-air-volume (CAV) system and the variable-air-volume (VAV) system. In the CAV system, supply air is distributed to spaces requiring ventilation and conditioned air through a series of ducts with inlet vents in the ceiling. Return air is drawn from the conditioned space through ceiling outlets into a large return air plenum. Dampers can be operated to control the amount of intake and exhaust air and the percentage of air that is recirculated. Temperature of the occupied space is controlled by regulating the temperature of the supply air. Proper operation of dampers/damper systems is critical for supplying sufficient ventilation air.

Unlike the CAV system, the VAV system regulates temperature in the occupied space by varying the amount of air supplied to the space. This type of system is attractive because it has the potential for significant energy savings. Unfortunately, there are drawbacks to this type of system. First, it is often difficult to ensure that minimum outdoor air ventilation rates for each zone are being satisfied. Second, the system is designed/operated to fully close dampers when temperature setting are reached. Consequently, no ventilation is provided to that space, and complaints of poor indoor air quality frequently arise (Godish 1989).

Ventilation problems experienced in mechanically ventilated building are often the result of poor system operation and/or system maintenance. Typical problems are

- building operated with 100% recirculated air in an attempt to conserve energy

- low ventilation efficiencies so that only a small fraction of the ventilation supply air can thoroughly mix in the occupied space before being exhausted

- system imbalance resulting in some space receiving more ventilation air than others.

3.2   INTERACTION WITH POLLUTANTS

When infiltration and natural ventilation fresh air input rates to a building are restricted, either intentionally or inadvertently, concentrations of indoor air contaminants will increase. Increasing minimum ventilation rates has been one response to this problem. However, a recent study con- ducted in the Pacific Northwest by Bonneville in 38 commercial buildings found poor correlation between ventilation rates and pollution levels (Turk et al. 1986). Nevertheless, the investigators attributed this poor correlation at least in part to the low pollutant levels observed, which makes the regression techniques unstable. The same conclusion may not be true for other buildings, particularly those with higher levels of contamination. This conclusion is consistent with research done on residential buildings.

The investigators also concluded that buildings with air exchange rates less than the minimum recommended (i.e., ASHRAE Standard 62-1981) do not necessarily have indoor air quality problems. Variations in pollutant source strength (indoor/outdoor), building volumes, and other removal processes can govern pollutant levels.

3.3   STANDARDS

The use of outdoor air for fresh-air ventilation assumes that the levels of contaminants in the ambient air are relatively low and safe for building occupants. ASHRAE Standards 62-1973 and 62-1981 recommend that outdoor air used for ventilation meet ambient air quality standards or be treated so that it does. Unfortunately, this standard is infrequently adhered to (Godish 1989).

The American Society of Heating, Refrigerating, and Air-Conditioning Engineers, Inc., has established minimum mechanical ventilation rates to achieve acceptable indoor air quality, assuming there are no unusual pollutant sources in the building. The first ASHRAE ventilation standard was published in 1973, modified in 1981, and modified again in 1989. ASHRAE Standard 62-1973 specified a minimum ventilation rate for any application of 5 cfm per person and a recommended rate of 20 to 25 cfm per person. The revised standard (ASHRAE 62-1981) specified minimum outside air rates for smoking and nonsmoking areas. The minimum rate for nonsmoking areas was still 5 cfm, but 20 to 35 cfm for smoking areas, depending upon building type.

In 1989, ASHRAE revised the ventilation standard in response to studies showing that a mechanical ventilation rate of 5 cfm was not sufficient to maintain acceptable levels of indoor pollutants. The new standard (ASHRAE 62-1989) increased the minimum outside airflow rate from 5 cfm to 15 cfm per person and combined the smoking and nonsmoking rates into one recommendation. The purpose of this standard is "to specify minimum ventilation rates and indoor air quality that will be acceptable to human occupants and are intended to avoid adverse health effects" (ASHRAE 1989).

The outside airflow rates recommended in Standard 62-1981 and Standard 62-1989 are compared in Table 3.1.

3.4 CHARACTERIZATION OF BUILDING AIR EXCHANGE RATES

A number of studies have been conducted in the past several years to measure the overall air exchange rates in commercial buildings. These studies were conducted in occupied buildings with the HVAC systems in their normal mode of operation. The results of these studies are summarized below.

In a study conducted in the Pacific Northwest, overall average air exchange rate measurements were made in 38 commercial buildings (Turk et al. 1987). The buildings were in two distinctly different climate regions. Ages of the buildings ranged from 0.5 year to 90 years. The overall average air exchange rate was 1.5 ACH with a standard deviation of 0.87 ACH (58% of the mean) with a range of 0.3 ACH to 4.1 ACH. The investigators concluded that no statistically significant relationship existed between ventilation rates and

TABLE 3.1. Recommended Airflow Rates (cfm/person) Contained in ASHRAE
Standard 62-1981 and 62-1989

Building Area Type	Standard 62-1981 Nonsmoking	Smoking	Standard 62-1989
Office spaces	5	20	20
Retail stores	5	25	0.02-0.30 cfm/ft^2
Classrooms	5	25	15
Dining rooms	7	35	20
Hotel conference rooms	7	35	20
Office conference rooms	7	35	20
Ballrooms and discos	7	35	25
Spectator areas	7	35	15
Theater auditoriums	7	35	15
Transporting waiting rooms	7	35	15
Hospital patient rooms	7	35	25
Residences	10	10	0.35 ACH[a]
Bars/cocktail lounges	10	50	30
Beauty shops	20	35	25
Smoking lounges	--	--	60

(a) air changes per hour

building height, building age, and number of stories above grade, due in part
to the dominating influences of season, air-handling equipment differences,
and HVAC system operating policies. Unfortunately, ventilation rates in some
of these buildings may have been artificially high because of building
operator actions during the monitoring period. In each instance of low
ventilation rates (<0.5 ACH), it was found that the outside air dampers were
completely closed during the monitoring. In two of the buildings, the dampers
were closed for energy conservation measures to reduce the cooling load in the
summer. In one building, the system operators were not sure of the damper
location or the control mechanism (Turk et al. 1986).

Grot and Persily (1986) measured air infiltration and ventilation rates
between 0.28 ACH and 0.70 ACH in eight federal buildings using tracer gas

techniques. The results for each building are shown in Table 3.2. The building minimum air ventilation rate is calculated from the ASHRAE guidelines based on building occupancy levels, building type, and room type. Half of the buildings did not meet the minimum ventilation rates recommended by ASHRAE.

Seven of these eight buildings were new (less than 3 years old at the time of testing) and constructed to the U.S. federal energy guidelines of less than 630 MJ/m^2 per year of off-site energy and less than 1200 MJ/m^2 per year of off-site energy. The building in Fayetteville was 7 years old and built before the energy guidelines were in effect. In general, the investigators concluded that these eight buildings performed better than most existing federal office buildings.

In other studies, ventilation rates in another eight buildings studied by Grot and Persily (1986) ranged from 0.33 ACH to 1.04 ACH. Silberstein and Grot (1985) measured an air exchange rate of 0.9 ACH in a single building.

In the commercial sector, little data exists to characterize air exchange rates in older buildings or to make comparisons between energy-efficient buildings and typical buildings.

TABLE 3.2. Average Ventilation/Infiltration for Eight Federal Buildings

Location	Average ACH	Bldg Minimum ACH
Anchorage	0.28	0.26
Ann Arbor	0.70	0.47
Columbia	0.40	0.62
Fayetteville	0.33	0.32
Huron	0.20	0.13
Norfolk	0.52	0.62
Pittsfield	0.32	0.38
Springfield	0.50	0.55

4. Mitigation

Managing indoor air quality (IAQ) is a process that may begin in the earliest phases of a building's design, or not be attempted until a problem is identified. The earlier in the building's life cycle that IAQ is addressed, the more control options building managers will have to choose from.

4.1 BUILDING COMMISSIONING

As a quality assurance procedure, building commissioning is the process of verifying and documenting the performance of building systems so that they conform with design intent. ASHRAE defines building commissioning as follows:

Commissioning is the process of achieving, verifying, and documenting that the performance of the HVAC system meets the design intent
and owners' functional criteria and operational needs. The process
extends through all phases of a project, from concept through
occupancy (ASHRAE 1988).

Four key benefits of pursuing building commissioning are as follows:

- Design intent is documented so that building owners, contractors,
  and operators can calculate and verify energy loads, as well as
  determine if specified indoor air quality mitigation techniques were
  properly installed.

- Occupant comfort is ensured, which is important for acceptance of
  energy-efficient technologies and may reduce the perception of
  problems with indoor air quality.

- The proper installation and operation of controls and equipment is
  verified to maximize energy savings and to meet applicable health
  and safety standards and design intentions.

- A building baseline is established which allows for accurate
  analysis of performance. Without knowing precisely what equipment
  is installed, it is difficult to assess performance in terms of
  expected and actual energy consumption targets.

During the past decade, the thermal environment, which is controlled by the HVAC system, has been the greatest source of occupant complaints in buildings (Trueman 1989a). It has become apparent that operations staff have had increasing difficulty in identifying and correcting HVAC system problems, particularly in new buildings. In almost every case where investigators have

61

found physical causes for complaints, two factors were at work. Previously undiscovered design deficiencies were identified, and building operators did not understand the proper operations of their building (Trueman 1989a). These factors have also been implicated as contributing to poor indoor air quality (Rask and Lane 1989).

In addition to controlling the thermal and ventilation comfort experienced by building occupants, HVAC systems are typically the single largest consumer of energy in a commercial building. Furthermore, HVAC systems may serve as the sole removal mechanism for indoor air pollutants. Thus, design deficiencies and improper operation have enormous implications for energy efficiency as well as comfort, health, and productivity.

Commissioning could include the inspection of equipment for problems that may lead to poor indoor air quality, and also could include monitoring. If unacceptable indoor air pollutant levels are found during commissioning, recommendations for remediation may include the use of elevated temperatures to encourage vapor offgassing, along with high ventilation rates to flush pollutants to the outdoors (Girman et al. 1989). These steps, referred to as building bake-out, will be most easily implemented if pursued prior to building occupancy.

4.1.1 Costs

Specific costs of the commissioning process will vary by building. However, Levin has suggested that if building commissioning begins in the first stages of a project, there should be no additional costs (ASHRAE 1989b). Trueman (1989b) has calculated the costs and benefits of building commissioning and concluded that the process will result in modest costs, but benefits will outweigh these expenditures two to one. Benefits result from reduced energy costs, corrected deficiencies, and corrected environmental problems. Trueman concludes that commissioning will result in cost increases to the building designer, contractor, and owner. These costs on a typical 100,000-square foot building would amount to a total in the range of $15,000 to $35,000.

4.1.2 Commercial Availability

ASHRAE (1988) has adopted guidelines for building commissioning, and workshops are being held to train engineers and architects in this process. In addition, inspection forms and guidelines from groups other than ASHRAE have been established (for example, the United States Army Corps of Engineers 1988). However, building commissioning is in its infancy and a definitive guide has not yet been developed. A document soon to be published by the National Institute of Standards and Technology may improve this situation.

4.2 SOURCE AVOIDANCE AND CONTROL

Source control refers to methods of control applied at a specific source of pollution, rather than for an entire building or zone. This section addresses nonmechanical methods of source control. Mechanical techniques, such as local ventilation, are discussed in Section 4.3. Techniques for source avoidance and control include removal, substitution, encapsulation, and timing applications for off-hours. Environmental factors such as high humidity or temperatures that contribute to pollutant emissions can also be avoided. These techniques refer primarily to specific sources of pollutants. For additional information, refer to specific pollutants in Section 2.0.

Source avoidance and substitution is likely to work best in the design of new buildings. In existing buildings, the removal of a pollutant source can require renovation, which may be among the most expensive of control techniques. In new buildings, designers have the opportunity to select and integrate mitigation strategies, such as specifying low-emitting materials, conditioning materials before installation, isolating potential sources of pollutants, and using building commissioning activities (see Section 4.1). For example, new building designers may employ the following strategies:

- incorporating isolated rooms for smokers with local exhaust or air treatment

- isolating and incorporating local exhaust in areas with high sources of pollutants, such as kitchens, work rooms, print shops, and laboratories

- using prefilters to reduce the intake of combustion and biogenic particles along with specifying low-emitting building materials

- using hard surfaces in interior designs and minimizing open storage shelving to avoid dust accumulation and to minimize growth areas for microbials

- providing the lowest comfortable air temperature to minimize VOC offgassing

- avoiding high occupant density and providing worker control and privacy.

4.2.1 Costs

Costs will vary by measure. If renovation is required, costs may be high. However, if material substitution is employed as part of planned renovation or new construction, costs may be minimal.

4.2.2 Commercial Availability

Suggestions for evaluating materials and writing specifications have been published (Levin 1989b). However, these procedures are not widespread, and existing guidance is general. Few building materials and products have been tested as potential sources of indoor air pollutants, and a universal grading system has yet to be developed.

4.3 INDOOR AIR QUALITY MANAGEMENT

Sections 4.1 and 4.2 describe engineering techniques to control IAQ. However, many of these techniques require building owners and business managers to commit financial and time resources to design solutions. For example, building commissioning requires engineering expertise, and also requires the building owner to train building operators and give them the time to properly maintain and operate mechanical and structural systems. Source avoidance can also include management decisions in scheduling the application of potential pollutant control.

Other techniques are directly related to management decisions. The following approaches from Levin (1989a) are suggested control measures for buildings with complex cases of sick building syndrome:

- Increase worker control over lighting, ventilation, heating, cooling, and noise.

- Provide flexible work hours where feasible.

- Minimize exposure in stressful positions through job rotation or mandatory rest periods.

Experience from risk communication programs suggests that communication and education efforts may also contribute to pollution control. These efforts can be directed to building operators and custodians, who directly control mechanical systems or may introduce pollutant sources. These efforts also may be important for low- and mid-level managers, as well as other staff. Staff can be trained to recognize potential sources of pollutants and possible symptoms of exposure. Staff can provide input to identify problem areas within buildings and in determining appropriate comfort levels for the building.

4.4 BUILDING DIAGNOSTICS

According to the National Academy of Science (NAS), building diagnostics refers to practices used to assess the current performance and capability of a building and predict its likely performance in the future (NAS 1985). Although many of these techniques can be used for commissioning new buildings, the term often applies to activities in existing buildings.

Efforts are under way to develop standard protocols for conducting building investigations. These protocols are important because diverse disciplines tend to become involved in these investigations. Environmental scientists, chemists, industrial hygienists, architects, mechanical engineers, and public health officials may all become involved in analyzing sick buildings. Thus, it is important that a common language and process be developed to facilitate accuracy and understanding. A four-step process had been identified:

- knowing what to measure
- determining appropriate instrumentation
- interpreting results
- predicting building performance.

The American Society for Testing and Materials (ASTM) has prepared a book documenting several diagnostic techniques (Nagda and Harper 1989). One of these approaches, presented by Woods et al. (1989), outlines a three-phase approach to investigating buildings, which incorporates the four diagnostic principles listed above. The three phases are

- *consultation* - In this phase the objectives and scope of the investigation are defined, a preliminary hypothesis may be formulated, and preliminary recommendations may be presented to the client. Measurements during this phase consist of professional observations by the investigators. A preliminary determination of required subsequent measurements is also made.

- *qualitative diagnostics* - Engineering analysis techniques may be used to validate recommendations or hypothesis. Performance criteria for the various functional areas of the building will be defined. If health problems are suspected, analysis of system performance will be initiated, and measurements will be limited to those required to evaluate the system. If health problems are suspected, immediate medical attention will be recommended for the affected occupants.

- *quantitative diagnostics* - Objective measurements of the physical environment and subjective responses of the occupants will be obtained. If further investigation is needed to test a hypothesis or to validate the recommendations, quantitative measurements of airborne contaminants, bulk samples, and other environmental parameters will be acquired through a systematic format. A thorough quality assurance/quality control program will be implemented.

4.4.1 Costs

References to costs were not found in the literature.

4.4.2 Commercial Availability

National conferences and workshops are conducted on building diagnostics, and professional societies have developed guidelines for implementation. Presenters at these conferences represent public health agencies, engineering firms, mechanical equipment manufacturers, industrial hygienists, and scientists. It appears that services in building diagnoses are becoming more widespread, but are not common in most mechanical engineering and architectural firms.

4.5 MECHANICAL CONTROL OF INDOOR AIR POLLUTANTS

An engineered approach to control of indoor air contaminants often falls into one of two major categories: ventilation and/or air treatment.

4.5.1 Mechanical Ventilation

Mechanical ventilation systems provide a means of reducing the concentrations of all indoor source contaminants by introducing outside air. If improperly designed, however, a ventilation system can create additional problems such as

- introduction of outside air contaminants into the building (e.g., CO from an adjacent parking garage, recirculation of exhausted air from the same or another building into the inlet air grille)

- inadvertent circulation of contaminants generated at one location of the building to other locations in the building

- increased emanation rates of pollutant sources due to feedback effects.

Three different strategies are used to improve indoor air quality through ventilation, including ventilation by dilution, localized exhaust ventilation, and improved room air distribution.

Dilution ventilation reduces indoor air pollutants by introducing a certain amount of outdoor air and exhausting indoor air. The amount of ventilation required to reduce indoor pollutant concentration levels varies. Large amounts of ventilation can result in a substantial rise in energy costs, providing an upper bound on how much ventilation is used. Typically, designers choose the minimum amount of ventilation required by local standards to minimize energy consumption. However, some field measurements suggest that the mechanical ventilation rates in buildings often exceed design rates (Seton, Johnson & Odell, Inc. 1984), a discrepancy that could be the result of either poor construction/commissioning practices or the manual override of ventilation system settings by building operators.

Localized exhaust ventilation is one way to minimize transport of pollutants from an identified source to the general indoor air. Such exhausts are commonly found in bathrooms, kitchens, laboratories, and smoking areas. These systems can be far more cost-effective than central dilution ventilation.

Ventilation systems perform less than optimally when stratification or nonuniform mixing of air within spaces in the building causes some portion of the supply ventilation air to bypass the occupied space. This "short-circuiting" may occur when supply and return air room diffusers are both located in proximity to each other. To describe the overall effectiveness of a building's ventilation system with respect to contaminant removal, engineers frequently refer to a ventilation efficiency factor. One proposed definition for ventilation efficiency (see ASHRAE 62-1989) is "...The measure of the effectiveness of the ventilation air to provide a contaminant level in the occupied zone of a space divided into the contaminant level produced by a similar amount of ventilation air with perfect mixing in the space..." (McNall 1986). Designing for higher ventilation efficiency appears to be an effective way to improve indoor air quality without significant energy penalties, but is generally practical only in new building design.

4.5.2 Air Treatment

Besides ventilation, air treatment methods are available to mitigate two categories of indoor air pollutants: gaseous and particulate. These methods are commonly used separately or in series in air-handling units, depending on the needs of the building, but also may be applied between a local source and the environment (e.g., recirculating range hoods). Using these methods, the air-handling unit's fan horsepower must be increased because of increased air flow resistance. However, ventilation air quantities needed to provide an equivalent amount of air treatment may then be reduced, lowering building cooling/heating loads, and thus HVAC equipment operating costs. In most commercial buildings, ventilation rates required by local codes are used by designers, because it is more cost-effective to employ one or more of the air treatment devices described below than to increase ventilation rates.

Particulate Removal

Two types of particulate removal systems are predominantly used in commercial building ventilation applications: fibrous filters and electronic air cleaners.

Fibrous Filters. The most common means to remove particulates from the air is through fibrous filtration systems. Typically, filters are mats of fine fibers through which building air passes. Particulates in the air passing through the filter collect on the fibers. Particulate removal systems are rated on their efficiency (i.e., ability to remove particulates), dust-holding capacity, and airflow resistance. Factors affecting the efficiency of fibrous filters include average fiber diameter in the filter, filter thickness, filter packing density/porosity, airflow rate, and particle size.

Several different fibrous filters are commercially available, with a wide range of efficiencies and costs:

- Dry-type panel filters have low efficiencies, and are used typically to stop dust. These filters are applied in most commercial building ventilation systems as either the sole filtration device or as a prefilter for higher efficiency filters.

- Viscous media panel filters have a viscous oily material coated over coarse fibers. Particulates adhere to this material after impacting the fibers. Both viscous and dry panel filters have relatively low pressure drops, and thus impose little energy penalty because of increased fan horsepower requirements. Viscous filters have very low efficiency at removing dust common to indoor air, but are used instead to collect fabric dust and lint (Godish 1989).

- Renewable media filters mechanically supply new filter media whenever the previous filter media becomes excessively clogged. Also called an automatic roll type filter, this mechanism is triggered by a pressure drop sensor across the filter media that has been preset for a maximum allowable pressure drop. Filter media is either dry-type or viscous in these systems.

- Extended surface dry-type filters are considered high-efficiency relative to panel filters, and work through extending filter surface area. This is achieved by either pleating the filter medium, or by creating open "bags" of filter material into which particulates are trapped. By extending the surface in this manner, rather than by simply making the filter thicker, the air pressure drop across the filter is minimized. The most efficient commonly used filter, known as the high-efficiency particulate air (HEPA), captures in excess of 99.97% of particulates at diameters equal or greater than 0.3 μm (Godish 1989).

To ensure lower fan operating costs and continued high effectiveness, fibrous filters typically must be replaced or reconditioned when they become excessively clogged.

Electronic Air Cleaners.   Electrostatic air cleaners remove particulates from an air stream on the principle of electrostatic attraction of opposite charges.   Three designs are common: the ionizing plate type, the charged-media non-ionizing type, and the charged-media ionizing type.

In the majority of applications, ionizing plate electronic air cleaners are used. Pressure drop through these air cleaners is very low, and they typically have a fairly high efficiency (80% to 95%). Eventually, as buildup of particulates on the collection plates increases, the effectiveness of the air cleaner decreases. Thus, frequent cleaning of the plates is required to ensure continued high efficiency (Godish 1989).

Charged media non-ionizing air cleaners consist of dielectric filtering medium mats in a field of alternately charged and grounded support members. A strong electrostatic field in the dielectric polarizes particles passing through the air cleaner, causing the particles to be drawn to the mats. Like standard dry-type filter mats, these mats must be periodically replaced or reconditioned.

Charged media ionizing air cleaners are similar to ionizing plate air cleaners except that instead of collection plates, a charged-media filter is used.

Two important environmental concerns are associated with the use of electronic air cleaners (ASHRAE 1988):

- *Space charging* occurs when particulates pass through the ionizer but are not deposited on the collection plates. If sufficient charge develops, these particles are driven to the building's walls. Thus, if the electronic air cleaner is malfunctioning or of low efficiency, inside wall surfaces can quickly become dirtier than if no cleaning device were used. A good design practice when using these systems is to include a fibrous afterfilter to help capture charged particles that escape the collection plates.

- *Ozone production* also may become a problem in electronic air cleaning systems if continuous arcing or brush discharge occurs.

Gaseous Contaminant Removal

Air treatment for the control of the amount of gaseous contaminants in indoor air theoretically may be accomplished by a variety of means, including adsorption, absorption, and catalytic oxidation. However, in commercial building applications, the most widely used method is adsorption and, thus, is the focus of the following discussion.

Adsorption refers to a phenomenon that causes gases, vapors, or liquids contacting a surface to bind to some degree to the surface. Once adsorbed, the gas or vapor molecules remain on the surface as a liquid or semi-liquid.

The adsorbents commonly used for indoor air cleaning come as multiple-sized granules, packed into beds that are configured to minimize resistance to airflow while maximizing dwell time of the air in the adsorption media. These adsorbents commonly have high surface area/volume ratios, and typically comprise of vast labyrinths of submicroscopic pores and minute channels (Godish 1989). Among the most popular adsorbents are activated carbons, activated alumina, silica gel, and surface-active clays. Activated carbon has become the adsorbent of choice in the vast majority of commercial building applications (Turk 1983).

Activated carbons are produced in a variety of sizes and hardnesses (Godish 1989). Hardness is important because a harder carbon helps ensure greater structural integrity for the panels in which it is packed and minimizes the risk that partially crushed granules within the panel would allow contaminants to escape in bypassed air, reducing efficiency of the gas cleaner. Granule size directly impacts the efficiency of an activated carbon gas cleaner. Higher density packings are possible with smaller granules, which improves efficiency, but also results in a greater airflow resistance, affecting the economics of the design being considered.

Activated carbons can vary in their degree of adsorptivity, depending on the characteristics of the application, the carbon, and the contaminants (Godish 1989). For example, the smaller pore diameters in the carbon result in higher surface areas, which, in turn, promote more adsorptivity. Adsorptive capacity can also vary depending on the contaminant being removed from the airstream. Table 4.1 shows the variation in adsorptive capacities for several different carcinogenic vapors. Elevated airstream humidities (>50%) can reduce activated carbon's adsorption capacity, because carbon's affinity for water vapor increases as humidity increases. However, in the context of air cleaning in commercial building applications, this is a small concern, because the retention rate of water on activated carbon is quite low and, consequently, water is displaced by sorbed gases and vapors, reducing the carbon's adsorptive capacity for water vapor.

TABLE 4.1. Adsorption Capacities of Activated Carbon
            for 27 Carcinogenic Vapors at 20°C

Vapor	Adsorption capacity, g material/g carbon
Acetamide	0.494
Acrylonitrile	0.357
Benzene	0.409
Carbon tetrachloride	0.741
Chloroform	0.688
bis(chloromethyl)ether	0.608
Chloromethyl methyl ether	0.480
1,2-dibromo-3-chloropropane	0.992
1,1-dibromomethane	0.962
1,2-dibromomethane	1.020
1,2-dichloroethane	0.575
Diepoxy butane (meso)	0.510
1,1-dimethyl hydrazine	0.359
1,2-dimethyl hydrazine	0.375
Dimethyl sulfate	0.615
p-dioxane	0.475
Ethylenimine	0.354
Hydrazine	0.380
Methyl methane sulfonate	0.595
1-naphthylamine	0.585
2-naphthylamine	0.506
N-nitrosodiethylamine	0.442
N-nitrosomethylamine	0.458
B-propiolactone	0.508
Prophylenimine	0.361
Vinyl chloride	0.404
Urethane	0.450

Analogous to particulate air cleaning systems, after being exposed to a
contaminant airstream long enough, a carbon air cleaner eventually reaches a
saturated state and can no longer adsorb any more contaminants (Godish 1989).
After this "breakpoint," the carbon must be replaced.

Activated carbon air cleaner performance can also be enhanced through
impregnation of materials on the carbon surface that selectively react with
molecules from the gas stream. The products of this reaction are then
retained on the carbon surface. This process is called chemisorption. Common
activated carbon impregnants and the pollutants they react with are shown in
Table 4.2 (Turk 1983).

TABLE 4.2. Activated Carbon Adsorption Impregnations

Impregnant	Pollutant	Action
Bromine	Ethylene; other alkenes	Conversion to dibromide, which remains on carbon
Lead acetate	H_2S	Conversion to PbS
Phosphoric acid	NH_3 amines	Neutralization
Sodium silicate	HF	Conversion to fluorosilicates
Iodine	Mercury	Conversion to HgI_2
Sulfur	Mercury	Conversion to HgS
Sodium sulfite	Formaldehyde	Conversion to addition product
Sodium carbonate or bicarbonate	Acidic vapors	Neutralization
Oxides of Cu, Cr, V, etc.; noble metals (Pd, Pt)	Oxidizable gases, including reduced sulfur compounds such as H_2S and mercaptans	Catalysis of air oxidation

Although performance data is available to indicate adsorption capacity of activated carbon, actual case studies of adsorption/chemisorption air cleaning systems used in commercial building applications are limited. In these cases, judgments of effectiveness are frequently subjective and largely anecdotal (e.g., an occupant assessment of whether an activated carbon air cleaner has removed an objectionable odor in a room or building).

Costs for Air Treatment Systems

Cost estimates for typical in-duct air treatment systems for commercial HVAC applications are described in the *National Mechanical Estimator* (Ottaviano 1987). These estimates include 1987 prices for initial installation and annual operating costs for a typical 10,000 cfm system. The annual operating costs include replacement media, labor charges, and fan power costs that would be incurred because of the pressure drop across the air treatment media. Although not current pricing, these estimates provide a good means of comparing different air treatment systems.

For bag-type filter arrangements with efficiencies ranging from 38% to 40% or 93% to 97%, the corresponding initial costs range from $431 to $1023, and annual operating expenses are $562 and $3129, respectively. Thus, increasing efficiency can be equated with a corresponding steep increase in cost, indicating the importance of carefully analyzing the building's air treatment needs before deciding on a system.

Particulate control using an electrostatic precipitator along with either bag filters or automatic roll filters has higher first costs ($5755) than comparable efficiency system exclusively using fibrous filters ($1053), but lower subsequent annual operating costs ($1222 versus 2402).

Although not directly comparable to particulate removal systems, the activated carbon gas contaminant removal system appears to have fairly high first costs of $7521 (e.g., twice as expensive HEPA systems) and smaller annual operating costs ($1496).

5. Energy Conservation Measures and Indoor Air Quality—Implications

The concentration of a pollutant in the indoor environment is dependent on the following factors:

- pollutant source strength
- location of the source in the building or outdoors
- transport and/or mixing of the pollutants within the building
- environmental factors such as temperature, humidity, and synergistic interactions with other pollutants
- the amount and distribution of outside air mixed with the indoor air through infiltration or mechanical ventilation
- removal mechanisms, such as filtration, exhaust ventilation, exfiltration, and air cleaning devices.

The quality of indoor air is dependent on the complex interaction between these factors. To the extent that ECMs may affect a number of these factors, the relationship between ECMs and indoor air quality is difficult to predict. In general, ECMs may affect pollutant concentrations in one of two ways: by introducing a source of pollutants or by changing the rate at which pollutants are removed from (or introduced to) an indoor space.

Potential sources of indoor VOCs from energy conservation measures include caulking compounds and insulation. Of the two potential sources discussed, caulking emits the highest rates of VOCs, but these are likely to be quickly dissipated. Insulation materials emit pollutants at a lower rate, but have the potential of emitting them over a longer time period. Emissions from fibrous insulation may be related to how moist the material is (van der Wal et al. 1987).

The second way in which energy conservation measures may affect indoor VOC levels is by changing the rate at which pollutants are removed from a space. The most common removal mechanism involves the exchange of indoor and outdoor air in ventilation systems. If conservation measures involve reducing the quantities of outside air without including air-cleaning equipment, indoor levels of pollutants are likely to increase. On the other hand, if conservation measures such as economizers are employed, increasing quantities of outside air, indoor pollutant levels are likely to decrease. A more thorough discussion of ventilation is contained in Section 3 of this report.

In existing buildings, energy conservation can affect the sources and removal rates of indoor pollutants through inspections and improved mainte- nance of the building structure and the HVAC system. Perhaps the most direct positive effects will be on microbial growth in the HVAC system and on build- ing surfaces. For example, caulking around windows may eliminate the entry of wind driven rain, which fosters the growth of molds and fungi (Rask and Lane 1989). Inspections may also discover microbial growth areas in ductwork near cooling coils, or design deficiencies such as the placement of air intakes near sources of pollutants such as garbage storage areas or parking garages (Rask and Lane 1989; Morey 1988).

However, if building inspections result in reductions in ventilation rates, there is likely to be a negative effect on indoor air quality. A field study suggests that ventilation rates often exceed ASHRAE Standard 62-1981. Seton, Johnson & Odell, Inc. (1984) conclude that nominal ventilation rates based on actual occupancy are significantly higher that the design rates listed in Standard 62-1981. Turk et al. (1987) found that in a sample of 40 buildings, on average, ventilation rates ranged from 2 to 8 times the rates recommended for smoking areas in Standard 62-1981. The ventilation rates contained in Standard 62-1981 for smoking areas (as opposed to the ventilation rates listed for nonsmoking areas) are greater than the single ventilation rates listed in ASHRAE Standard 62-1989. Thus, Turk's findings suggest that in existing buildings, ventilation rates exceed both the new ASHRAE Standard 62-1989 and the old ASHRAE Standard 62-1981. If the ventilation rates are reduced to match either of the standards, there is likely to be a negative effect on indoor pollution concentrations.

Table 5.1 summarizes the potential impacts of ECMs on indoor air quality. For this discussion, the ECMs will have either a direct or an indirect impact on the indoor air quality. A direct impact is one that directly increases (decreases) the source strength by increasing (decreasing) the amount of source material or its emanation rate. An indirect impact is

TABLE 5.1.   List of Candidate Energy Conservation Measures and Associated IAQ Impacts

Energy Conservation Measure	RSP Fibers	RSP Particles	Biogenic	Radon	Combustion Gases	VOCs
Building Envelope						
Install wall, ceiling, roof, floor, foundation and slab insulation	D+	nc	nc	nc	nc	D+
Reduce heat/cool loads from infiltration by caulking and weatherstripping	I+	I±	I±	I+	I±	D+ I+
Install storm windows/doors	I±	I±	I-	I+	I±	I+
Enclose loading docks with shelters and seals	I±	I±	I±	I+	I±	I+
Install vestibules to reduce infiltration/exfiltration	I±	I±	I±	I+	I±	I+
Seal vertical shafts to reduce in/exfiltration	I±	I±	I±	I+	I±	I+
Heating/Air Conditioning						
Install automatic condenser cleaning	u	u	D-	u	u	D+
Replace air-cooled condenser with cooling tower	nc	nc	D+	nc	nc	nc
Install spot cooling	nc	nc	D+	nc	nc	nc
Install earth cooling tubes	nc	nc	D+	nc	nc	nc

TABLE 5.1. (contd)

Energy Conservation Measure	RSP Fibers	RSP Particles	Biogenic	Radon	Combustion Gases	VOCs
Install roof spray systems	nc	nc	D+	nc	nc	nc
Install economizer (air-side)	u	u	u	u	u	I-
Install warm-up cycle controls/optimum start	u	u	u	u	u	I~
Install automatic setback/setup thermostats	u	u	u	u	u	I+
Insulate ducts	D+	nc	I-	nc	nc	D+
Ventilation						
Install CO_2-controlled ventilation	u	u	u	u	I-	nc
Install CO-controlled covered parking ventilation	u	u	u	u	D-	nc
Automatically reduce ventilation during unoccupied periods	I±	I±	I±	I+	I±	I+
Reduce minimum outside air	u	u	u	u	u	I+
Recirculate exhaust air using activated carbon filters	D+	I-	I-	I-	I-	I~
Install vortex hoods for restaurants	nc	I-	nc	nc	I-	I-

TABLE 5.1. (contd)

Energy Conservation Measure	RSP Fibers	RSP Particles	Biogenic	Radon	Combustion Gases	VOCs
Employ evaporative cooling of outdoor air	nc	nc	D+	nc	nc	nc
Employ desiccant dehumidification	u	u	u	u	u	nc
Reduce energy consumption for fans by reducing air flow rates and resistance to air flow	nc	nc	nc	nc	nc	nc
Install high efficiency fans with larger ductwork	u	u	u	u	u	I~
Install dual speed fans	u	u	u	u	u	I~
Install attic ventilation	nc	nc	nc	I+	nc	I-
Install low leakage dampers	u	u	u	u	u	I-
Install ceiling fans	u	u	u	u	u	nc
Install outside air reset control	u	u	u	u	u	nc

KEY   D = direct effect
      I = indirect effect
      + = increase pollutant levels
      - = decrease pollutant levels
      ± = either increase or decrease depending
          on whether indoor or outdoor source
      nc = no change
      u = unsure

one that modifies or in some way increases (decreases) the pollutant concentration in the indoor environment through increasing (decreasing) the dilution rate or removal rate. Materials used in the ECM itself may also be a direct source of an indoor pollutant. These are indicated in Table 5.1 by the D+ symbol. They fall into two broad categories--those that increase amounts of thermal insulation material and those that modify the HVAC system and increase the available growth media for biogenic material. For instance, adding thermal insulation to the HVAC system ducts will increase the amount of fiber material in the building and increase the potential for release of RSP fibers to the air, distributed by the HVAC system.

6. References

ASHRAE. 1972. *ASHRAE Standard 62-1972 - Ventilation for Acceptable Indoor Air Quality.* American Society of Heating, Refrigerating, and Air-Conditioning Engineers, Inc., Atlanta, Georgia.

ASHRAE. 1981. *ASHRAE Standard 62-1981 - Ventilation for Acceptable Indoor Air Quality.* American Society of Heating, Refrigerating, and Air-Conditioning Engineers, Inc., Atlanta, Georgia.

ASHRAE. 1988. *ASHRAE Guideline - HVAC Systems Commissioning.* American Society of Heating, Refrigerating, and Air-Conditioning Engineers, Inc., Atlanta, Georgia.

ASHRAE. 1989a. *ASHRAE Standard 62-1989 - Ventilation for Acceptable Indoor Air Quality.* American Society of Heating, Refrigerating, and Air-Conditioning Engineers, Inc., Atlanta, Georgia.

ASHRAE. 1989b. "Excerpts from Summary Workshop." *The Human Equation: Health and Comfort*, Proceedings of the ASHRAE/SOEG Conference IAQ 89, April 12-20, 1989, San Diego, California. American Society of Heating, Refrigerating, and Air-Conditioning Engineers, Inc., Atlanta, Georgia.

Baechler, M.C., H. Ozkaynak, J.D. Spengler, L.A. Wallace, and W.C. Nelson. 1989. Assessing Indoor Exposure to Volatile Organic Compounds Released from Paint Using the NASA Data Base." In *Proceedings of the Annual Meeting of the Air and Waste Management Association*, Anaheim, California, June 20-22. Air and Waste Management Association, Pittsburgh, Pennsylvania.

Berglund, B., I. Johansson, and T. Lindvall. 1987. "Volatile Organic Compounds from Building Materials in a Simulated Chamber Study." In *Indoor Air '87: Proceedings of the 4th International Conference on Indoor Air Quality and Climate.* Institute for Water, Soil and Air Hygiene, Berlin.

Bonneville Power Administration. 1982. *Environmental Assessment of Energy Conservation Opportunities in Commercial-Sector Facilities in the Pacific Northwest.* DOE/EA-0187, Portland, Oregon.

Bonneville Power Administration. 1984. *Final Environmental Impact Statement: The Expanded Residential Weatherization Program.* DOE/EIS-0095F, Portland, Oregon.

Bonneville Power Administration. 1988. *Final Environmental Impact Statement on New Energy-Efficient Homes Programs: Assessing Indooor Air Quality Options.* DOE/EIS-0127F, Portland, Oregon.

Borrazzo, J.E., J.F. Osborn, R.C. Fortmann, and C.I. Davidson. 1985. *Airborne Concentrations of CO, NO$_x$, and Particulates in a Modern Townhouse-- Model Predictions vs. Experimental Results*. Departments of Civil Engineering, Engineering & Public Policy, and Biomedical Engineering, Carnegie-Mellon University, Pittsburgh, Pennsylvania.

Brundage, J.F., R.M Scott, W.M. Lednar, D.W. Smith, and R.N. Miller. 1988. "Building-Associated Risk of Febrile Acute Respiratory Diseases in Army Trainees." *Journal of the American Medical Association* 259(14):2108-2112.

Burge, H.A. 1985. "Indoor Sources for Airborne Microbes," *Indoor Air and Human Health*, Proceedings of the Seventh Life Sciences Symposium, Knoxville, Tennessee, October 29-31, 1984, Garnage and Kaye, eds., pp. 139-148.

Burge, P.S. 1990. "Building Sickness - A Medical Approach to the Causes." *In Indoor Air 90: Proceedings of the Fifth International Conference on Indoor Air Quality and Climate*. Canada Mortgage and Housing Corporation, Toronto.

Carey, J.M., M. Hager, P. King, and S. Zuckerman. January 7, 1985. "Beware Sick-Building Syndrome." *Newsweek*.

Crouse W.E., M.S. Ireland, J.M. Johnson, R.M. Striegel, Jr., C.S. Williard, R.M. DePinto, G.B. Oldaker III, and R.L. McBride. 1988. "Results from a Survey of Environmental Tobacco Smoke (ETS) in Restaurants." In *Transactions of an International Specialty Conference: Combustion Processes and the Quality of the Indoor Environment*, ed. J.P. Harper, pp. 214-219. September 1988, Niagara Falls, New York.

Fanger, P.O. 1989. "The New Comfort Equation for Indoor Air Quality." In The Human Equation: Health and Comfort, Proceedings of the ASHRAE/SEOG Conference IAQ 89, April 12-20, 1989, San Diego, California. American Society of Heating, Refrigerating, and Air-Conditioning Engineers, Inc., Atlanta, Georgia.

Fanger, P.O. 1990. "New Principles for a Future Ventilation Standard." *In Indoor Air 90: Proceedings of the Fifth International Conference on Indoor Air Quality and Climate*. Canada Mortgage and Housing Corporation, Toronto.

Flachsbart, P.G, and D. E. Brown. 1985. "Merchant Exposure to CO from Motor Vehicle Exhaust at Honolulu's Ala Moana Shopping Center." In *Proceedings of the 78th Annual Meeting of the Air Pollution Control Association*, June 16-21, 1985, Detroit, Michigan. Air Pollution Control Association, Pittsburgh, Pennsylvania.

Gamboa, R.R., B.P. Gallagher, and K.R. Mathews. 1988. "Data on Glass Fiber Contribution to the Supply Airstream from Fiberglas Duct Liner and Fiberglas Duct Board." In *IAQ 88: Engineering Solutions to Indoor Air Problems*, pp. 25-33. American Society of Heating, Refrigerating, and Air-Conditioning Engineers, Inc., Atlanta, Georgia.

Girman J., L. Alvantis, and G. Kulasingam. 1987. "Bake Out of an Office Building." In *Indoor Air "87: Proceedings of the 4th International Conference on Indoor Air Quality and Climate.* Institute for Water, Soil and Air Hygiene, Berlin.

Girman, J.R., L. Alevantis, M. Petreas, and L. Webber. 1989. "The Bake-Out of an Office Building: A Case Study." *Environment International* 15:449-453.

Godish, T. 1989. *Indoor Air Pollution Control.* Lewis Publishers, Inc., Chelsea, Michigan.

Grot, R.A., and A.K. Persily. 1986. "Measured Air Infiltration and Ventilation Rates in Eight Large Office Buildings." In *Measured Air Leakage of Buildings, ASTM STP 904,* eds. H.R. Trechel and P.L. Lagus, pp. 151-183. American Society for Testing and Materials, Philadelphia.

Hawthorne, A.R., and T.G. Mathews. 1985. "Formaldehyde: An Important Indoor Pollutant." In *Proceedings: Indoor Air Quality Seminar--Implications for Electric Utitlity Conservation Programs,* EPRI EA/EM-3824, pp. 8.1-8.7. Electric Power Research Institute, Palo Alto, California.

Jungers, R.H., and L.S. Sheldon. 1987. "Characterization of Volatile Organic Chemicals in Public Access Buildings." *In Indoor Air '87: Proceedings of the 4th International Conference on Indoor Air Quality and Climate.* Institute for Water, Soil and Air Hygiene, Berlin.

Kandarjian, L. 1988. "Federal Policy Options for Indoor Air Pollution from Combustion Appliances." In *Transactions of an International Specialty Conference: Combustion Processes and the Quality of the Indoor Environment,* ed. J.P. Harper, pp. 310-332. September 1988, Niagara Falls, New York.

Levin H. 1989a. "Sick Building Syndrome: Review and Exploration of Causation Hpotheses and Control Methods." In *The Human Equation: Health and Comfort,* Proceedings of the ASHRAE/SOEG Conference IAQ 89, April 12-20, 1989, San Diego, California. American Society of Heating, Refrigerating, and Air-Conditioning Engineers, Inc., Atlanta, Georgia.

Levin H. ed. 1989b. "Specifying Interior Materials." *Indoor Air Quality Update* 2(4):5-11.

McNall, P.E. 1986. "Strategies: Today and Tomorrow." *ASHRAE Journal* 28(7):37-40.

McNall, P.E. 1988. "Control Technology and IAQ Problems." In *Engineering Solutions to Indoor Air Problems,* Proceedings of the ASHRAE Conference IAQ 88, April 11-13, 1988, Atlanta Georgia. American Society of Heating, Refrigerating, and Air-Conditioning Engineers, Inc., Atlanta, Georgia.

Melius, J., K. Wallingford, J. Carpenter, and R. Keenslyside. 1984. "Indoor Air Quality: The NIOSH Experience." *Am Conf Ind Hyg Report* 10:3-7.

Miele, P.F. 1989. "Formaldehyde Emissions from Bonded Fiberglass Insulation Products." In *The Human Equation: Health and Comfort*, Proceedings of the ASHRAE/SOEG Conference IAQ 89, April 12-20, 1989, San Diego, California. American Society of Heating, Refrigerating, and Air-Conditioning Engineers, Inc., Atlanta, Georgia.

Miksch, R.R., C.D. Hollowell, and H.E. Schmidt. 1982. *Trace Organic Chemical Contaminants in Office Spaces*. LBL-12561, Lawrence Berkeley Laboratory, Berkeley, California.

Molhave, L. 1982. "Indoor Air Pollution Due To Organic Gases and Vapours of Solvents in Building Materials." *Environment International* 8:117-127.

Molhave, L. 1985. "Volatile Organic Compounds as Indoor Air Pollutants." In *Indoor Air and Human Health, Proceedings of the Seventh Life Sciences Symposium*, eds. R.B. Gammage and S.V. Kaye, pp. 403-414. October 29-31, 1984, Knoxville, Tennessee. Lewis Publishers, Inc, Chelsea, Michigan.

Molhave, L., B. Bach, and O.F. Pedersen. 1985. "Human Reactions During Controlled Exposures to Low Concentrations of Organic Gases and Vapours Known as Normal Indoor Air Pollutants." *In Indoor Air. Vol. 3: Sensory and Hyperactivity Reactions to Sick Buildings - Proceedings of the International Conference (3rd) on Indoor Air Quality and Climate*, pp. 431-436. Swedish Council for Building Research, Stockholm.

Molhave, L. 1990. "Volatile Organic Compounds, Indoor Air Quality and Health." *In Indoor Air 90: Proceedings of the Fifth International Conference on Indoor Air Quality and Climate*. Canada Mortgage and Housing Corporation, Toronto.

Morey, P.R. 1988. "Microorganisms in Buildings and HVAC Systems: A Summary of 21 Environmental Studies." In *Engineering Solutions to Indoor Air Problems*, Proceedings of the ASHRAE Conference IAQ 88, April 11-13, 1988, Atlanta Georgia. American Society of Heating, Refrigerating, and Air-Conditioning Engineers, Inc., Atlanta, Georgia.

Morey, P.R., and B.A. Jenkins. 1989. "What are Typical Concentrations of Fungi, Total Volatile Organic Compounds, and Nitrogen Dioxide in an Office Environment?" In *The Human Equation: Health and Comfort*, Proceedings of the ASHRAE/SOEG Conference IAQ 89, April 12-20, 1989, San Diego, California. American Society of Heating, Refrigerating, and Air-Conditioning Engineers, Inc., Atlanta, Georgia.

Mueller Associates, Inc., Argonne National Laboratory, and Brookhaven National Laboratory. 1987. *Indoor Air Quality Environmental Information Handbook: Building System Characteristics*. DOE/EV/10450-H1, U.S. Department of Energy, Washington, D.C.

Mueller Associates, Inc., SYSCON Corporation, and Brookhaven National Laboratory. 1985. *Indoor Air Quality Environmental Information Handbook: Combustion Sources*. DOE/EV/10450-1, U.S. Department of Energy, Washington, D.C.

Nagda N.L., and J.P. Harper, eds. 1989. *Design and Protocol for Monitoring Indoor Air Quality*. The American Society for Testing and Materials, Philadelphia.

National Academy of Science (NAS). 1985. *Building Diagnostics: A Conceptual Framework*. National Academy Press, Washington, D.C.

Ottaviano, V.B. 1987. *National Mechanical Estimator*. Ottaviano Technical Services, Inc., and Fairmont Press, Atlanta, Georgia.

Ozkaynak, H., and P.B. Ryan. 1987. "Sources and Emission Rates of Organic Chemical Vapors in Homes and Buildings." *In Indoor Air '87: Proceedings of the 4th International Conference on Indoor Air Quality and Climate*. Institute for Water, Soil and Air Hygiene, Berlin.

Rask, D.R., and C.A. Lane. 1989. "Resolution of the Sick Building Syndrome: Part II, Maintenance." In *The Human Equation: Health and Comfort*, Proceedings of the ASHRAE/SOEG Conference IAQ 89, April 12-20, 1989, San Diego, California. American Society of Heating, Refrigerating, and Air-Conditioning Engineers, Inc., Atlanta, Georgia.

Seifert, B, D. Ullrich, and R. Nagel. 1989. "Volatile Organic Compounds from Carpeting" in *Man and His Ecosystem; Proceedings of the 8th World Clean Air Congress*, 11-15 September 1989, The Hague. Elsevier Science Publishers, Amsterdam.

Seton, Johnson & Odell, Inc. 1984. *Ventilation in Commercial Buildings*. Prepared for the Office of Conservation, Bonneville Power Administration, Portland, Oregon.

Sexton, K., J.D. Spengler, and R.D. Treitman. 1984. "Effects of Residential Wood Combustion on Indoor Air Quality: A Case Study in Waterbury, Vermont." *Atmospheric Environment*.

Sheldon, L.S., R.W. Handy, T.D. Hartwell, R.W. Whitmore, H.S. Zelon, and E.D. Pellizzari. 1988a. *Indoor Air Quality in Public Buildings*, Vol 1. EPA/600/S6-88/009a. U.S. Environmental Protection Agency, Washington, D.C.

Sheldon, L., H. Zelon, J. Sickles, C. Eaton and T. Hartwell. 1988b. *Indoor Air Quality in Public Buildings*, Vol. 2. U.S. Environmental Protection Agency, Washington, D.C.

Silberstein, S., and R.A. Grot. 1985. "Air Exchange Rates Measurements of the National Archive Building." *ASHRAE Transactions* 91(2A).

Skov, P., and O. Valbjorn. 1987. "The Sick Building Syndrome in the Office Environment: The Danish Town Hall Study, Indoor Air '87." In *Proceedings of the 4th International Conference on Indooor Air Quality and Climate*, Berlin.

Spengler, J.D., and K. Sexton. 1983. "Indoor Air Pollution: A Public Health Perspective." *Science* 221(4605):9-17.

Spengler, J.D. and M.A. Cohen, 1985. "Emissions from Indoor Combustion Sources". *Indoor Air and Human Health*, Proceedings of the Seventh Life Sciences Symposium, Knoxville, TN, October 29-31, 1984, Garnage and Kaye, eds., pp. 261-278.

Sterling, T.D., C.W. Collett, and J.A. Ross. 1988. "Levels of Environmental Tobacco Smoking Under Different Conditions of Ventilation and Smoking Regulations." In *Transactions of an International Specialty Conference: Combustion Processes and the Quality of the Indoor Environment*, ed. J.P. Harper, pp. 223-235. September 1988, Niagara Falls, New York.

Sterling, E.M., and T.D. Sterling. 1984. "Baseline Data: Health and Comfort in Modern Office Buildings." *In Proceedings of the 5th Air Infiltration Center Conference: The Implementation and Effectiveness of Air Infiltration Standards in Buildings*. International Energy Agency, The Air Infiltration Center, Berkshire, England.

Tichenor, B.A., and M.A. Mason. 1988. "Organic Emissions from Consumer Products and Building Materials to the Indoor Environment." *JAPCA* 38:264-268.

Tichenor, B.A., L.E. Sparks, M.D. Jackson, Z. Guo, and S.A. Rasor. 1990. "The Effect of Wood Finishing Products on Indoor Air Quality." Presented at *Measurement of Toxic and Related Air Pollutants*, Raleigh, North Carolina, 30 April - 4 May 1990. Sponsored by the U.S. Environmental Protection Agency and the Air and Waste Management Association.

Trueman, C.S. 1989a. Untitled presentation to *Dimension '89 - A Seminar for the Building Design Professional*, March 15-17, 1989.

Trueman, C.S. 1989b. "Commissioning: An Owner's Approach for Effective Operations." *ASHRAE Transactions 1989*. American Society of Heating, Refrigerating, and Air-Conditioning Engineers, Inc., Atlanta, Georgia.

Turk, A. 1983. "Gaseous Air Cleaning Can Help Maintain Tolerable Indoor Air Quality Limits." ASHRAE Journal, pp. 35-37.

Turk, B.H., J.T. Brown, K. Geisling-Sobotka, D.A. Froehlich, D.T. Grimsrud, J. Harrison, and K.L. Revzan. 1986. "Indoor Air Quality Measurements in 38 Pacific Northwest Commercial Buildings." In *Proceedings of the 79th Annual Meeting of the Air Pollution Control Association*, Minneapolis, Minnesota. Air Pollution Control Association, Pittsburgh, Pennsylvania.

Turk, B.H, J.T. Brown, K. Geisling-Sobotko, D.A. Froehlich, G.T. Grimsrud, J. Harrison, J.F. Koonce, R.J. Prill, and K.L. Revzan. 1987. *Indoor Air Quality and Ventilation Measurements in 38 Pacific Northwest Commercial Buildings.* Bonneville Power Administration, Portland, Oregon.

Turner, S. 1988. "Environmental Tobacco Smoke and Smoking Policy." In *Transactions of an International Specialty Conference: Combustion Processes and the Quality of the Indoor Environment*, ed. J.P. Harper, pp. 236-247.

U.S. Army Corps of Engineers (USACOE). 1988. Acceptance Test Procedure for Air Supply and Distribution Systems.

U.S. Environmental Protection Agency (EPA). 1988. "Indoor Air Facts No. 4: Sick Buildings." Washington, D.C.

U.S. Department of Energy (DOE). 1986. *Environmental Assessment in Support of Proposed Interim Energy Conservation Standards for New Commercial and Multifamily Highrise Residential Buildings.* DOE/CE-0166, Washington, D.C.

Valbjorn, O., and P. Skov. 1987. "Influence of Indoor Climate on the Sick Building Syndrome, Indoor Air." In *Proceedings of the 4th International Conference on Indoor Air Quality and Climate.* Institute for Water, Soil and Air Hygiene, Berlin.

van der Wal, J.F., A.M.M. Moons, and R. Steenlage. 1987. "Thermal Insulation as a Source of Air Pollution." In *Indoor Air '87: Proceedings of the 4th International Conference on Indoor Air Quality and Climate.* Institute for Water, Soil and Air Hygiene, Berlin.

Wallace, L. 1985. "Part Five: Overview." In *Indoor Air and Human Health, Proceedings of the Seventh Life Sciences Symposium*, eds. R.B. Gammage and S.V. Kaye, pp. 331-333. October 29-31, 1984, Knoxville, Tennessee. Lewis Publishers, Inc, Chelsea, Michigan.

Wallace, L. 1987. "Emissions of Volatile Organic Compounds from Building Materials and Consumer Products." In *Indoor Air '87: Proceedings of the 4th International Conference on Indoor Air Quality and Climate.* Institute for Water, Soil and Air Hygiene, Berlin.

Wallace L., R. Jungers, L. Sheldon, and E. Pellizzari. 1987. "Volatile Organic Chemicals in 10 Public-Access Buildings." In *Indoor Air '87: Proceedings of the 4th International Conference on Indoor Air Quality and Climate.* Institute for Water, Soil and Air Hygiene, Berlin.

White, J.B., J.C. Reaves, P.C. Reist, and L.S. Mann. 1988. "A Data Base on the Sources of Air Pollution Emissions." In *Engineering Solutions to Indoor Air Problems*, Proceedings of the ASHRAE Conference IAQ 88, April 11-13, 1988, Atlanta Georgia. American Society of Heating, Refrigerating, and Air-Conditioning Engineers, Inc., Atlanta, Georgia.

Woods, J.E., P.R. Morey, and D.R. Rask. 1989. "Indoor Air Quality Diagnostics: Qualitative and Quantitative Procedures to Improve Environmental Conditions." In *Design and Protocol for Monitoring Indoor Air Quality*, eds. N.L. Nagda and J.P. Harper. The American Society for Testing and Materials, Philadelphia.

Part II

Health Effects

The information in Part II is from *Health Effects Associated with Energy Conservation Measures in Commercial Buildings—Volume 2: Review of the Literature,* prepared by R.D. Stenner and M.C. Baechler of Pacific Northwest Laboratory (operated by Battelle Memorial Institute for the U.S. Department of Energy) for Bonneville Power Administration under an agreement with the U.S. Department of Energy, September 1990.

DISCLAIMER

Summary

Indoor air quality can be impacted by hundreds of different chemicals. More than 900 different organic compounds alone have been identified in indoor air. Health effects that could arise from exposure to individual pollutants or mixtures of pollutants cover the full range of acute and chronic effects, including largely reversible responses, such as rashes and irritations, to the irreversible toxic and carcinogenic effects. These indoor contaminants are emitted from a large variety of materials and substances that are widespread components of everyday life.

Pacific Northwest Laboratory conducted a search of the peer-reviewed literature on health effects associated with indoor air contaminants for the Bonneville Power Administration to aid the agency in the preparation of environmental documents.

Results are reported in two volumes. Volume 1 summarizes the results of the search of the peer-reviewed literature on health effects associated with a selected list of indoor air contaminants. In addition, the report discusses potential health effects of polychlorinated biphenyls and chlorofluorocarbons. All references to the literature reviewed are found in Volume 2. Volume 2 provides detailed information from the literature reviewed, summarizes potential health effects, reports health hazard ratings, and discusses quantitative estimates of carcinogenic risk in humans and animals.

Contaminants discussed in this report are those that

- have been measured in the indoor air of a public building or

- have been measured (significant concentrations) in test situations simulating indoor air quality (as presented in the referenced literature) and

- have a significant hazard rating.

This report reflects information currently available on the health effects associated with each of the selected contaminants. This information was obtained by conducting a comprehensive literature search on each of the selected contaminants. In addition, three active, professionally maintained and peer-reviewed on-line chemical databases were used to locate the latest health effects information on each contaminant. The on-line databases used were the U.S. Environmental Protection Agency's Integrated Risk Information System (IRIS), the National Library of Medicine's Hazardous Substance Data Bank (HSDB), and the National Institute for Occupational Safety and Health's Registry of Toxic Effects of Chemicals (RTECS).

It is recognized that the chemicals addressed in this report likely represent only a small number of the host of potential hazardous contaminants that may be found inside public buildings. In this report, public buildings are considered to be those commercial buildings that are open to the public.

1. Introduction

The energy crisis of 1973 and its consequences focused attention on the problem of indoor air pollution. Among the measures taken to conserve energy have been the "tightening" of existing buildings and the design and construction of "tighter" new buildings. The tightening of buildings saves energy by activities such as reducing drafts around doors, windows, openings for pipes and ducts, through electrical outlets, cracks, and other openings. When people living and working in these tightened buildings began to complain in increased numbers about headaches and general malaise, investigations revealed accumulations of potentially toxic or carcinogenic chemicals because these buildings were insufficiently ventilated. In tightened buildings, indoor air concentrations of these chemicals can exceed outdoor values as well as national air-quality standards. Because the average person spends roughly 90% of his/her life indoors, the importance of indoor air quality for human health is readily apparent.

This report presents the best available information on the health effects associated with each of the selected contaminants. This information was obtained by conducting a comprehensive literature search on each of the selected chemicals. In addition, three active, professionally maintained and peer-reviewed on-line chemical databases were used to locate the latest health effects information on each contaminant. The on-line databases used were the U.S. Environmental Protection Agency (EPA) Integrated Risk Information System (IRIS), the National Library of Medicine's Hazardous Substance Data Bank (HSDB), and the National Institute for Occupational Safety and Health's Registry of Toxic Effects of Chemicals (RTECS).

This report is presented in two volumes. Volume 1 summarizes the results of a search of the peer-reviewed literature on health effects associated with a selected list of indoor air contaminants. In addition, the report discusses potential health effects of polychlorinated biphenyls and chlorofluorocarbons. Volume 2 provides detailed information from the literature reviewed, summarizes potential health-effects, reports health hazard ratings, and discusses quantitative estimates of carcinogenic risk in humans and animals.

This report discusses only those contaminants that have been measured in the indoor air of a public building or measured (significant concentrations) in test situations simulating indoor air quality (as presented in the referenced literature) and that have a significant hazard rating. It is recognized that the chemicals addressed in this report are likely to represent only a small number of the host of potential hazardous contaminants that may be found inside public buildings. In this report public buildings are considered to be those commercial buildings that are open to the public.

1.1 DEFINITION OF TERMS

A glossary of definitions is appropriate to establish a common set of terms for discussing health effects of contaminants as they relate to the performance of risk assessments and health assessment evaluations. Definitions for the terms presented in this section were taken from the IRIS database for risk assessment information maintained by the EPA (IRIS 1990). These definitions are not intended to be all-encompassing, nor should they be construed to be "official" definitions. It is assumed that the user has some familiarity with risk assessment and health science. For terms that are not included in this glossary, the user should refer to standard health science, biostatistical, and medical textbooks and dictionaries.

Acceptable Daily Intake -- An estimate of the daily exposure dose that is likely to be without deleterious effect even if continued exposure occurs over a lifetime.

Acute exposure -- One dose or multiple doses occurring within a short time (24 hours or less).

Acute hazard or toxicity -- see Health hazard.

Added risk -- The difference between the cancer incidence under the exposure condition and the background incidence in the absence of exposure; AR = P(d) - P(0).

Anecdotal data -- Data based on descriptions of individual cases rather than on controlled studies.

Anemia -- A condition marked by significant decreases in hemoglobin concentration and in the number of circulating red blood cells.

Angina pectoris -- Constricting chest pain which may be accompanied by pain radiating down the arms, up into the jaw, or to other sites.

Anoxic stress -- Oxygen starvation.

Ataxia -- Loss of coordination.

Attributable risk -- The difference between risk of exhibiting a certain adverse effect in the presence of a toxic substance and that risk in the absence of the substance.

Benign -- Not malignant; remaining localized.

Bioassay -- The determination of the potency (bioactivity) or concentration of a test substance by noting its effects in live animals or in isolated organ preparations, as compared with the effect of a standard preparation.

Bioavailability -- The degree to which a drug or other substance becomes available to the target tissue after administration or exposure.

Bronchiolalveolitis -- Inflammation in the bronchi and alveoli.

Bronchospasm -- Temporary narrowing of the bronchi due to violent, involuntary contraction of the smooth muscle of the bronchi.

Carcinogen -- An agent capable of inducing a cancer response.

Carcinogenesis -- The origin or production of cancer, very likely a series of steps. The carcinogenic event so modifies the genome and/or other molecular control mechanisms in the target cells that these can give rise to a population of altered cells.

Case-control study -- An epidemiologic study that looks back in time at the exposure history of individuals who have the health effect (cases) and at a group who do not (controls), to ascertain whether they differ in proportion exposed to the chemical under investigation.

Chronic effect -- An effect that is manifest after some time has elapsed from initial exposure. See also Health hazard.

Chronic exposure -- Multiple exposures occurring over an extended period of time, or a significant fraction of the animal's or the individual's lifetime.

Chronic hazard or toxicity -- see Health hazard.

Chronic study -- A toxicity study designed to measure the (toxic) effects of chronic exposure to a chemical.

Cohort study -- An epidemiologic study that observes subjects in differently exposed groups and compares the incidence of symptoms. Although ordinarily prospective in nature, such a study is sometimes carried out retrospectively, using historical data.

Confounder -- A condition or variable that may be a factor in producing the same response as the agent under study. The effects of such factors may be discerned through careful design and analysis.

Conjunctivitis -- Irritation of the conjunctiva, the mucous membrane lining the inner surface of the eyelids and covering the front part of the eyeballs.

Control group -- A group of subjects observed in the absence of agent exposure or, in the instance of a case/control study, in the absence of an adverse response.

Core grade(s) -- Quality ratings, based on standard evaluation criteria established by the Office of Pesticide Programs, given to toxicological studies after submission by registrants.

Critical effect -- The first adverse effect, or its known precursor, that occurs as the dose rate increases.

Cyanosis -- A bluish color in the skin and mucous membranes due to deficient levels of oxygen in the blood.

Developmental toxicity -- The study of adverse effects on the developing organism (including death, structural abnormality, altered growth, or functional deficiency) resulting from exposure prior to conception (in either parent), during prenatal development, or postnatally up to the time of sexual maturation.

Dose-response relationship -- A relationship between the amount of an agent (either administered, absorbed, or believed to be effective) and changes in certain aspects of the biological system (usually toxic effects), apparently in response to that agent.

Dyspnea -- Difficult or labored breathing.

Emphysema -- Swelling of the air sacs or the tissue connecting them in the lungs, accompanied by atrophy of the tissues and impaired breathing.

Endpoint -- A response measure in a toxicity study.

Erythema -- Skin redness.

Estimated exposure dose (EED) -- The measured or calculated dose to which humans are likely to be exposed considering exposure by all sources and routes.

Excess lifetime risk -- The additional or extra risk incurred over the lifetime of an individual by exposure to a toxic substance.

Extra risk -- The added risk to that portion of the population that is not included in measurement of background tumor rate; ER(d) = [P(d) - P(0)]/[1-P(0)].

Extrapolation -- An estimation of a numerical value of an empirical (measured) function at a point outside the range of data which were used to calibrate the function. The quantitative risk estimates for carcinogens are generally low-dose extrapolations based on observations made at higher doses. Generally, one has a measured dose and measured effect.

Fetotoxic -- Toxic to the fetus.

Gamma multi-hit model -- A dose-response model of the form

P(d) = integral from 0 to d of {[a**k][s**(k-1)][exp(-as)]/G(u)}ds

where G(u) = integral from 0 to infinity of [s**(u-1)][exp(-s)]ds
      P(d) = the probability of cancer from a dose rate d
         k = the number of hits necessary to induce the tumor
         a = a constant

when k = 1, see the One-hit model.

Genotoxic -- Toxic to the genetic material within a person's cells [i.e., in the genes formed from the substance deoxyribonucleic acid (DNA), which make up the chromosomes in a person's cells].

Health Advisory -- An estimate of acceptable drinking water levels for a chemical substance based on health effects information; a Health Advisory is not a legally enforceable federal standard, but serves as technical guidance to assist federal, state, and local officials.

Health hazard (types of) --

1. Acute toxicity -- The older term used to describe immediate toxicity. Its former use was associated with toxic effects that were severe (e.g., mortality) in contrast to the term "subacute toxicity" that was associated with toxic effects that were less severe. The term "acute toxicity" is often confused with that of acute exposure.

2. Allergic reaction -- Adverse reaction to a chemical resulting from previous sensitization to that chemical or to a structurally similar one.

3. Chronic toxicity -- The older term used to describe delayed toxicity. However, the term "chronic toxicity" also refers to effects that persist over a long period of time whether or not they occur immediately or are delayed. The term "chronic toxicity" is often confused with that of chronic exposure.

4. Idiosyncratic reaction -- A genetically determined abnormal reactivity to a chemical.

5. Immediate versus delayed toxicity -- Immediate effects occur or develop rapidly after a single administration of a substance, while delayed effects are those that occur after the lapse of some time. These effects have also been referred to as acute and chronic, respectively.

6. Reversible versus irreversible toxicity -- Reversible toxic effects are those that can be repaired, usually by a specific tissue's ability to regenerate or mend itself after chemical exposure, while reversible toxic effects are those that cannot be repaired.

7. Local versus systemic toxicity -- Local effects refer to those that occur at the site of first contact between the biological system and the toxicant; systemic effects are those that are elicited after absorption and distribution of the toxicant from its entry point to a distant site.

Hematopoietic -- Having to do with the blood-forming elements.

Hepatic -- Having to do with the liver.

Hemoptysis -- Discharge of blood from the larynx, trachea, bronchi or lungs.

Homologue -- One of a series of organic compounds that differ from each other by a CH_2, such as the methane series C_nH_{2n+2}, in which there is a similarity between the compounds in the series and a graded change of their properties.

Human equivalent dose -- The human dose of an agent that is believed to induce the same magnitude of toxic effect as that which the known animal dose has induced.

Hydranencephaly -- A congenital anomaly in which the brain is not fully developed.

Hyperchromic anemia -- Any of several blood disorders in which red blood cells show an increase in hemoglobin and a reduction in number.

Hypoxia -- Oxygen deficiency to a reduction of the blood's oxygen-carrying capacity.

Incidence -- The number of new cases of a disease within a specified period of time.

Incidence rate -- The ratio of the number of new cases over a period of time to the population at risk.

Individual risk -- The probability that an individual person will experience an adverse effect. This is identical to population risk unless specific population subgroups can be identified that have different (higher or lower) risks.

Initiation -- The ability of an agent to induce a change in a tissue which leads to the induction of tumors after a second agent, called a promoter, is administered to the tissue repeatedly. See also Promoter.

Interspecies dose conversion -- The process of extrapolating from animal doses to equivalent human doses.

Latency period -- The time between the initial induction of a health effect and the manifestation (or detection) of the health effect; crudely estimated as the time (or some fraction of the time) from first exposure to detection of the effect.

Limited evidence -- According to the EPA's Guidelines for Carcinogen Risk Assessment, limited evidence is a collection of facts and accepted scientific inferences which suggests that the agent may be causing an effect, but this suggestion is not strong enough to be considered established fact.

Linearized multistage procedure -- The modified form of the multistage model (see Multistage model). The constant q1 is forced to be positive (>0) in the estimation algorithm and is also the slope of the dose-response curve at low doses. The upper confidence limit of q1 (called q1*) is called the slope factor.

Logit model -- A dose-response model of the form

$$P(d) = 1/[1 + \exp -(a + b \log d)]$$

where P(d) is the probability of toxic effects from a continuous dose rate d, and a and b are constants.

Lowest-observed-adverse-effect level (LOAEL) -- The lowest exposure level at which there are statistically or biologically significant increases in frequency or severity of adverse effects between the exposed population and its appropriate control group.

Lowest-effect level (LEL) -- Same as LOAEL.

Lymphocytes -- The principal cell type of the lymphs, comprising 20% to 30% of white blood cells.

Malignant -- Tending to become progressively worse and to result in death if not treated; having the properties of anaplasia, invasiveness, and metastasis.

Margin of Exposure (MOE) -- The ratio of the no observed adverse effect level (NOAEL) to the estimated exposure dose (EED).

Margin of Safety (MOS) -- The older term used to describe the margin of exposure.

Mesothelioma -- A tumor of the pleura (lining of the lung and chest cavity).

Metastasis -- The transfer of disease from one organ or part to another not directly connected with it; adj., metastatic.

Methemoglobinemia -- The presence of methemoglobin (oxidized hemoglobin) in the blood, which reduces the oxygen-carrying capacity of the blood.

Model -- A mathematical function with parameters that can be adjusted so that the function closely describes a set of empirical data. A "mathematical" or "mechanistic" model is usually based on biological or physical mechanisms, and has model parameters that have real-world interpretation. In contrast, "statistical" or "empirical" models are curve-fitting to data where the math function used is selected for its numerical properties. Extrapolation from mechanistic models (e.g., pharmacokinetic equations) usually carries higher confidence than extrapolation using empirical models (e.g., logit).

Modifying factor (MF) -- An uncertainty factor that is greater than zero and less than or equal to 10; the magnitude of the MF depends upon the professional assessment of scientific uncertainties of the study and database not explicitly treated with the standard uncertainty factors (e.g., the completeness of the overall database and the number of species tested); the default value for the MF is 1.

Multistage model -- A dose-response model often expressed in the form

$$P(d) = 1 - exp \{-[q(0) + q(1)d + q(2)d**2 + ... + q(k)d**k]\}$$

where $P(d)$ is the probability of cancer from a continuous dose rate d, the $q(i)$ are the constants, and k is the number of dose groups (or, if less, k is the number of biological stages believed to be required in the carcinogenesis process). Under the multistage model, it is assumed that cancer is initiated by cell mutations in a finite series of steps. A one-stage model is equivalent to a one-hit model.

Mutagenic -- Causing cell mutation, chromosome alternation, bacterial mutations, an DNA damage.

Nasopharyngeal -- Having to do with the nose and pharynx.

Neoplasm -- An aberrant new growth of abnormal cells or tissues; tumor.

Neurological sequelae -- A neurologic (nervous system) condition that results from or follows a disease, disorder or injury (i.e., complications of a neurologic injury).

No data -- According to the EPA Guidelines for Carcinogen Risk Assessment, "no data" describes a category of human and animal evidence in which no studies are available to permit one to draw conclusions as to the induction of a carcinogenic effect.

No evidence of carcinogenicity -- According to the EPA Guidelines for Carcinogen Risk Assessment, a situation in which there is no increased incidence of neoplasms in at least two well-designed and well-conducted animal studies of adequate power and dose in different species.

No-observed-adverse-effect level (NOAEL) -- An exposure level at which there are no statistically or biologically significant increases in the frequency or severity of adverse effects between the exposed population and its appropriate control; some effects may be produced at this level, but they are not considered as adverse, nor precursors to adverse effects. In an experiment with several NOAELs, the regulatory focus is primarily on the highest one, leading to the common usage of the term NOAEL as the highest exposure without adverse effect.

No-observed-effect level (NOEL) -- An exposure level at which there are no statistically or biologically significant increases in the frequency or severity of any effect between the exposed population and its appropriate control.

One-hit model -- A dose-response model of the form

$$P(d) = a - \exp(-b\ d)$$

where $P(d)$ is the probability of cancer from a continuous dose rate d, and b is a constant. The one-hit model is based on the concept that a tumor can be induced after a single susceptible target or receptor has been exposed to a single effective dose unit of a substance.

Organoleptic -- Affecting or involving a sense organ as of taste, smell, or sight.

Pheochromocytoma -- A tumor of the sympathetic nervous system composed principally of chromaffin cells; found most often in the adrenal medulla.

Pneumoconiosis -- Any lung disease caused by dust inhalation.

Principal study -- The study that contributes most significantly to the qualitative and quantitative risk assessment.

Probit model -- A dose-response model of the form

$$P(d) = 0.4\{\text{integral from minus infinity to } [\log(d - u)]/s \text{ of } [\exp-(y**2)/2]dy\}$$

where $P(d)$ is the probability of cancer from a continuous dose rate d, and u and s are constants.

Promoter -- In studies of skin cancer in mice, an agent that results in an increase in cancer induction when administered after the animal has been exposed to an initiator, which is generally given at a dose that would not result in tumor induction if given alone. A cocarcinogen differs from a promoter in that it is administered at the same time as the initiator.

Cocarcinogens and promoters do not usually induce tumors when administered separately. Complete carcinogens act as both initiator and promoter. Some known promoters also have weak tumorigenic activity, and some also are initiators. Carcinogens may act as promoters in some tissue sites and as initiators in others.

Proportionate mortality ratio (PMR) -- The number of deaths from a specific cause and in a specific period of time per 100 deaths in the same time period.

Prospective study -- A study in which subjects are followed forward in time from initiation of the study. This is often called a longitudinal or cohort study.

Pulmonary edema -- Fluid in the lungs.

q1* -- Upper bound on the slope of the low-dose linearized multistage procedure.

Reference Dose (RfD) -- An estimate (with uncertainty spanning perhaps an order of magnitude) of a daily exposure to the human population (including sensitive subgroups) that is likely to be without appreciable risk of deleterious effects during a lifetime.

Registration (of a pesticide) -- Under FIFRA and its amendments, new pesticide products cannot be sold unless they are registered with the EPA. Registration involves a comprehensive evaluation of risks and benefits based on all relevant data.

Regulatory dose (RgD) -- The daily exposure to the human population reflected in the final risk management decision; it is entirely possible and appropriate that a chemical with a specific RfD may be regulated under different statutes and situations through the use of different RgDs.

Relative risk (sometimes referred to as risk ratio) -- The ratio of incidence or risk among exposed individuals to incidence or risk among nonexposed individuals.

Renal -- Having to do with the kidneys.

Reportable quantity -- The quantity of a hazardous substance is considered reportable under CERCLA. Reportable quantities are 1) one pound, or 2) for selected substances, an amount established by regulation either under CERCLA or under Section 311 of the Clean Water Act. Quantities are measured over a 24-hour period.

Rhinitis -- Inflammation of the nasal mucous membrane.

Risk -- The probability of injury, disease, or death under specific circumstances. In quantitative terms, risk is expressed in values ranging from zero (representing the certainty that harm will not occur) to 1 (representing the certainty that harm will occur). The following are examples showing the manner in which risk is expressed: E-4 = a risk of 1/10,000; E-5 = a risk of 1/100,000; E-6 = a risk of 1/1,000,000. Similarly, 1.3E-3 = a risk of 1.3/1000 = 1/770; 8E-3 = a risk of 1/125; and 1.2E-5 = a risk of 1/83,000.

Risk assessment -- The determination of the kind and degree of hazard posed by an agent, the extent to which a particular group of people has been or may be exposed to the agent, and the present or potential health risk that exists due to the agent.

Risk management -- A decisionmaking process that entails considerations of political, social, economic, and engineering information with risk-related information to develop, analyze, and compare regulatory options and to select the appropriate regulatory response to a potential chronic health hazard.

Safety factor -- See Uncertainty factor.

Short-term exposure -- Multiple or continuous exposures occurring over a week or so.

Slope Factor -- The slope of the dose-response curve in the low-dose region. When low-dose linearity cannot be assumed, the slope factor is the slope of the straight line from 0 dose (and 0 excess risk) to the dose at 1% excess risk. An upper bound on this slope is usually used instead of the slope itself. The units of the slope factor are usually expressed as 1/(mg/kg-day).

Standardized mortality ratio (SMR) -- The ratio of observed deaths to expected deaths.

Subchronic exposure -- Multiple or continuous exposures occurring usually over 3 months.

Subchronic study -- A toxicity study designed to measure effects from subchronic exposure to a chemical.

Sufficient evidence -- According to the EPA Guidelines for Carcinogen Risk Assessment, sufficient evidence is a collection of facts and scientific references which is definitive enough to establish that the adverse effect is caused by the agent in question.

Superfund -- Federal authority, established by the Comprehensive Environmental Response, Compensation, and Liability Act (CERCLA) in 1980, to respond directly to releases or threatened releases of hazardous substances that may endanger health or welfare.

Supporting studies -- Those studies that contain information that is useful for providing insight and support for the conclusions.

Systemic effects -- Systemic effects are those that require absorption and distribution of the toxicant to a site distant from its entry point, at which point effects are produced. Most chemicals that produce systemic toxicity do not cause a similar degree of toxicity in all organs, but usually demonstrate major toxicity to one or two organs. These are referred to as the target organs of toxicity for that chemical.

Systemic toxicity -- See Systemic effects.

Tachypnea -- An abnormally fast rate of breathing.

Target organ of toxicity -- See Systemic effects.

Teratogen -- Causing rectal abnormalities, skeletal and visceral malformations, and functional/behavioral deviations.

Threshold -- The dose or exposure below which a significant adverse effect is not expected. Carcinogens are thought to be non-threshold chemicals, to which no exposure can be presumed to be without some risk of adverse effect.

Threshold limit values (TLVs) -- Recommended guidelines for occupational exposure to airborne contaminants published by the American Conference of Governmental Industrial Hygienists (ACGIH). The TLVs represent the average concentration (in mg/m^3) for an 8-hour workday and a 40-hour work week to which nearly all workers may be repeatedly exposed, day after day, without adverse effect.

Time-weighted average -- An allowable exposure concentration averaged over a normal 8-h workday or 40-h workweek.

Toxicity value -- A numerical expression of a substance's dose-response relationship that is used in risk assessments. The most common toxicity values used in EPA risk assessments are reference doses (for noncarcinogenic effects) and slope factors (for carcinogenic effects).

Tumor progression -- The sequence of changes in which a tumor develops from a microscopic lesion to a malignant stage.

Uncertainty factor -- One of several, generally ten-fold factors, used in operationally deriving the Reference Dose (RfD) from experimental data. UFs are intended to account for 1) the variation in sensitivity among the members of the human population; 2) the uncertainty in extrapolating animal data to the case of humans; 3) the uncertainty in extrapolating from data obtained in a study that is of less-than-lifetime exposure; and 4) the uncertainty in using LOAEL data rather than NOAEL data.

Unit Risk -- The upper-bound excess lifetime cancer risk estimated to result from continuous exposure to an agent at a concentration of 1 $\mu g/L$ in water, or 1 $\mu g/m^3$ in air.

Upper bound -- An estimate of the plausible upper limit to the true value of the quantity. This is usually not a statistical confidence limit.

Weibull model -- A dose-response model of the form

$$P(d) = 1 - \exp[-b(d**m)]$$

where P(d) is the probability of cancer due to continuous dose rate d, and b and m are constants.

Weight-of-evidence for carcinogenicity -- The extent to which the available biomedical data support the hypothesis that a substance causes cancer in humans.

1.2 THE NEPA PROCESS

The Bonneville Power Administration (Bonneville) has taken a leading role among federal agencies in assessing the environmental impacts of indoor air pollutants and designing appropriate program responses. These efforts have included extensive research programs into residential ventilation and indoor pollution characterization and monitoring. Bonneville has also prepared three environmental documents under the National Environmental Policy Act (NEPA) that focused on indoor air quality issues (BPA 1982, 1984, and 1988).

One of these documents, the 1982 *Environmental Assessment of Energy Conservation Opportunities in Commercial-Sector Facilities in the Pacific Northwest*, supported conservation programs in existing commercial buildings. Bonneville is now planning the implementation of aggressive commercial conservation acquisition programs in both new and existing buildings. Because of changing information, and because of the change in scope of the programs being designed, Bonneville is now reevaluating the potential environmental effects of conservation activities in commercial buildings. Pacific Northwest Laboratory[a] (PNL) prepared this review of information about health effects to aid Bonneville in its assessment of potential environmental effects.

1.3 SELECTION OF POLLUTANTS FOR HEALTH EFFECT ANALYSIS

The primary environmental concern identified for new energy-efficient buildings is the potential effect of increased levels of indoor pollutants on

---

(a) Pacific Northwest Laboratory is operated by Battelle Memorial Institute for the U.S. Department of Energy.

public health. A host of organic and other pollutants potentially exist in building materials and could be released to the indoor air of these public buildings. To maintain a degree of realism in this general health effects assessment report on the potential contaminants associated with the indoor air quality of public buildings, only those contaminants that have been measured in the indoor air of a public building or measured (significant concentrations) in test situations simulating indoor air quality situations were examined in this health effects study. An additional criterion applied was that the hazard rating (Sax and Lewis 1989) needed to be high enough for the contaminant to be considered a significant hazard. The contaminants found in building materials but not found in sampling efforts associated with public building were not included in the list to study. However, some of these contaminants will be briefly discussed in the section on Other Potential Contaminants. Table 1.1 contains a list of the contaminants examined in this contaminant health effects study.

This list was derived after reviewing current literature articles regarding studies of indoor air quality in public buildings, the National Aeronautics and Space Administration (NASA) test for potential contaminants to the atmosphere of spacecraft, and studies of building materials and consumer products (Baechler et al. 1989; Jungers and Sheldon 1987; Wallace et al. 1984; Wallace et al. 1987a and 1987b; Girman et al. 1982; Tichenor and Mason 1988; Molhave 1982; Krause et al. 1987; Wallace 1987; Wallace 1986; and Tichenor and Mason 1986).

1.4 STUDY METHODOLOGY

Upon establishing the list of contaminants to study, a comprehensive literature search was conducted to located the latest information on health effects associated with each of the chemicals on the list. In addition, three active, professionally maintained and peer-reviewed, on-line chemical information databases were used to located the latest health effect information on each contaminant on the list. The three databases accessed were 1) Integrated Risk Information System (IRIS), 2) Hazardous Substance Data Bank (HSDB), and 3) Registry of Toxic Effects of Chemicals (RTECS).

TABLE 1.1.   Contaminants Examined in Health Effects Study

Benzene	Formaldehyde
Benzo(A)Pyrene	Methylethylketone
Biogenic particles	Nitrogen oxides
n-butylacetate	Octane
Carbitol	Octene
Carbon dioxide	α-Pinene
Carbon monoxide	Polychlorinated
Carbon tetrachloride	Biphenyls
Chlorofluorocarbons	Propylbenzene
Chloroform	Radon/Radon daughters
Decane	Styrene
Dichlorobenzene	Synthetic fibers (Fiber
Dichloromethane	Glass & Cellulose)
Dodecane	Tetrachloroethylene
Environmental tobacco	Toluene
Smoke	1,1,1-Trichloroethane
2-Ethoxyethylacetate	Trichloroethylene
Ethylbenzene	1,2,4-Trimethylbenzene
Ethylene glycol	Undecane
	Xylene

IRIS is an on-line database of chemical-specific risk information prepared by the EPA in order to serve as guidance for EPA risk assessments.   IRIS
currently contains summaries of hazards, dose-response assessments and
reference citations on approximately 400 chemicals.   The database provides
information for carcinogenicity assessment such as human and animal
carcinogenicity data, weight of evidence for carcinogenicity, oral and
inhalation slope factors, unit risks and concentrations at specified risk
levels for chronic exposures to carcinogens.   It provides risk assessment
information such as oral and inhalation reference doses (RfDs) for chronic
noncarcinogenic health effects.   The health assessment information contained
in IRIS is peer-reviewed by an interdisciplinary review group of EPA
scientists and health professionals (Sidhu 1989).

   The HSDB is a factual, nonbibliographic databank focusing upon the toxicology of potentially hazardous chemicals.   The HSDB is managed by the
National Library of Medicine, MEDLARS Management Section.   It is enhanced with
data from such related areas as emergency handling procedures, environmental
fate, human exposure, detection methods, and regulatory requirements.   Data is

derived from a core set of standard texts and monographs, government documents, technical reports, and the primary journal literature. HSDB contains
complete references for all data sources utilized. The HSDB file is fully
peer reviewed by the Scientific Review Panel (SRP), a committee of experts
drawn from the major subject disciplines within the databanks's scope. HSDB
is organized by chemical record, with approximately 4,200 chemical substance
records contained in the file (NIH 1989).

RTECS is an on-line file containing toxic effects data on over
96,000 chemicals. Both acute and chronic effects are covered, including data
on skin/eye irritation, carcinogenicity, mutagenicity, and reproductive consequences. Selected federal regulatory requirements and exposure levels are
also presented. References are available for all data. Toxicology and
carcinogenic reviews, where available on any given chemical, are cited. Some
reviews may come from the International Agency for Research on Cancer (IARC)
Monograph Series or may be other general review articles from the open scientific literature. RTECS is built and maintained by the National Institute for
Occupational Safety and Health (NIOSH) (NIH 1989).

The information obtained in the literature search and active databases
were then combined to prepare this health effects assessment report. For the
most part, an emphasis was placed on general health effects and the health
effects associated with the airborne exposure pathway, since this is likely to
be the most dominant pathway of exposure for individuals residing/working in
public buildings. The health effect information available for each contaminant varied widely. Some of the contaminants, such as PCBs and formaldehyde,
had a vast amount of health effect-related information available. Other
contaminants, such as α-pinene and octene, had little or no health effect
information available. It was considered important to use only well peer-
reviewed and documented (by reliable sources) information on the health
effects associated with the selected contaminants. Thus, although for some
contaminants the information provided may be limited, the health effect
information provided in this report should be considered the best available at
this time.

2. Indoor Air Quality

Indoor air quality can be potentially impacted by conceivably hundreds of different chemicals. The health effects that could potentially arise from exposure to individual pollutants or mixtures of pollutants cover the full range of acute and chronic effects, including largely reversible responses, such as rashes and irritations, as well as irreversible toxic and carcinogenic effects. These indoor contaminants are emitted from a large variety of materials and substances that are widespread components of everyday life.

Asbestos, combustion gases, formaldehyde, and radon have been the focus of indoor air quality studies for some time. However, the understanding of complex mixtures of these and other commonly available chemicals is only in its infancy. As an example, environmental tobacco smoke contains over 3,800 different chemicals (NRC 1986). Similarly, the large number of volatile organics that emanate from a variety of consumer products, building materials, and furnishings are in the very early stages of quantitative study to relate specific organics to specific material sources (Tichenor 1989). In addition, a variety of biogenic contaminants are potentially available in public buildings that can affect occupants in many different ways.

This section discusses the health effects associated with the contaminants selected for this study, based on the criteria discussed earlier.

2.1 THE SICK BUILDING SYNDROME AND BUILDING-RELATED ILLNESS

A building is characterized as a sick building when its occupants complain of health and comfort problems that can be related to working or being in the building. The problems associated with sick buildings are sick building syndrome and building-related illness (EPA 1988). These terms generally apply to problems related to indoor air pollution. They are not normally used to characterize buildings where complaints stem solely from inadequate temperature or humidity control. A World Health Organization Committee estimates that up to 30% of new and remodeled buildings may have such problems.

108

A building is considered to manifest the sick building syndrome when a substantial percentage of the building occupants complain of symptoms associated with acute discomfort (e.g., headaches; eye, nose and throat irritations; dry cough; dry or itchy skin; dizziness and nausea; difficulty in concentrating; fatigue; sensitivity to odors), the cause of the symptoms is not known, and most of the complainants report relief upon leaving the building.

When occupant exposure to indoor contaminants results in a clinically defined illness, disease or infirmity, the building is said to manifest building-related illness. The characterization of building-related illness is by complaints of symptoms such as cough, chest tightness, fever, and muscle aches which can be associated with illness; the cause(s) of the symptoms are believed to be exposure to indoor pollutants; and complainants usually require prolonged recovery after leaving the building.

It is important to note that it is normal for a certain percentage of individuals in a building to experience one or more of the symptoms of sick buildings, and many of the complaints may be the result of illnesses contracted outside of the building or from stress-related circumstances. However, studies do show that such symptoms may be caused or heightened by indoor air contaminants.

Potential causes of sick buildings are inadequate ventilation, pollutants emitted inside the building, contaminants brought in from outside sources, and biological contamination. These causes usually act in combination and often supplement other occupant complaints, such as inadequate temperature, humidity, or lighting.

2.2 VOLATILE ORGANIC COMPOUNDS

Considerable emphasis is being placed on the wide range of volatile organic compounds that may be found in indoor air. There are numerous sources of indoor volatile organic compounds. More than 350 different organic compounds have been identified in concentrations over 0.001 parts per million (ppm) in indoor air (Wallace et al. 1984). This large and ever-growing list of organic compounds make it difficult to associate health and well-being problems associated with specific compounds. Actual exposures are to the various combinations of mixtures of these chemicals.

Organic compounds are part of almost all materials and products, such as construction materials, furnishings, combustion fuels, consumer products, and pesticides. A large variety of organic compounds are produced from combustion of cooking and heating fuels, tobacco, and human metabolism. Examples of organic compound types and their sources are listed in Table 2.1.

2.2.1 Benzene

Benzene is a clear, colorless liquid with a density of 0.8794 and a flash point of 12°F. Its molecular formula is C_6H_6, and its molecular weight is 78.12. It has a vapor pressure of 100 mm at 26.1°C. It has a hazard rating of 3 (Sax and Lewis 1989). The half-life of benzene in air is 6 days (EPA 1984a). Some of the more common synonyms for benzene are benzol, coal naphtha, cyclohexatriene, phene, phenyl hydride, polystream, and pyrobenzol. The threshold limit value (TLV), on a time weighted average (TWA) basis, for benzene is 10 ppm (ACGIH 1988).

2.2.1.1 Summary

The weight-of-evidence classification for benzene is "A; human carcino-gen." Several studies of increased incidence of nonlymphocytic leukemia from occupational exposure, increased incidence of neoplasia in rats and mice exposed by inhalation and gavage, and some supporting data form the basis for this classification (IRIS 1990). Both gavage and inhalation exposure of rodents to benzene have resulted in development of neoplasia.

Benzene and its metabolites inhibit both nuclear and mitochondrial replication and transcription. Deoxyriboneucleic acid (DNA) synthesis was inhibited in hemopoietic cells from mice exposed to a single dose of 3000 ppm benzene.

The inhalation slope factor is 2.9E-2/mg/kg/day. The inhalation unit risk is 8.3E-6/μg/m^3. The extrapolation method used was the one-hit model using pooled data. Table 2.2 shows air concentrations at specified risk levels (IRIS 1990).

TABLE 2.1.   Examples of Organic Compounds and Sources

Pollutant type	Examples	Example Indoor Sources
Aliphatic and Oxygenated Aliphatic Hydrocarbons	α-pinene, n-decane, n-undecane, n-dodecane, propane, butane, n-butylacetate, ethoxyethylacetate	Cooking and heating fuels, aerosol propellants, cleaning compounds, paints, carpet, moldings, particle board, refrigerants, lubricants, flavoring agents, perfume base
Halogenated Hydrocarbons	Methyl chloroform, methylene chloride, polychlorinated biphenylss, 1,1,1-trichloroethane, trichloroethylene, tetrachloroethylene, chlorobenzene, dichlorobenzene	Aerosol propellants, fumigants, pesticides, refrigerants, adhesives, caulk, paint, linoleum tile, carpet, latex paint, and degreasing, dewaxing, and dry cleaning solvents
Aromatic Hydrocarbons	Xylenes, ethylbenzenes, trimethylbenzenes, ethyltoluenes, propylbenzenes, benzene, styrene, toluene	Paints, varnishes, glues, enamels, lacquers, cleaners, adhesives, molding, insulation, linoleum tile, carpet
Alcohols	Ethanol, methanol	Lacquers, varnishes, polish removers, adhesives
Ketones	Acetone, diethyl ketone, methyl ethyl ketone	Lacquers, varnishes, polish removers, adhesives, cleaners
Aldehydes	Formaldehyde, nonanal	Fungicides, germicides, disinfectants, artificial and permanent-press textiles, paper, particle boards, cosmetics, flavoring agents

2.2.1.2 Health Effects

Noncarcinogenic Effects - Benzene is on EPA's list of chemicals to be reviewed by an EPA work group for determining the Reference Dose for chronic oral exposure (RfDo) and for chronic inhalation exposure (RfDi).

TABLE 2.2. Benzene Air Concentrations at Specified Risk Levels

Risk Level	Concentration
E-4 (1 in 10,000)	$1E+1\ \mu g/m^3$
E-5 (1 in 100,000)	$1E+0\ \mu g/m^3$
E-6 (1 in 1,000,000)	$1E-1\ \mu g/m^3$

Effect on Replication and Transcription - Benzene and its metabolites inhibit both nuclear and mitochondrial replication and transcription. Deoxyribonucleic acid (DNA) synthesis was inhibited in hemopoietic cells from mice exposed to a single dose of 3000 ppm benzene. It was also inhibited in mouse L5178YS lymphoma cells after their exposure to the metabolites, but not to benzene, which is not bioactivated in these cells. The most potent inhibitor was p-benzoquinone, followed by hydroquinone, 1,2,4-benzenetriol, catechol, and phenol, all of which are metabolites of benzene, at concentrations that were not cytotoxic. Inhibition correlated with ease of oxidation. This correlation suggests that the oxidation of phenol or one of its metabolites produces the ultimate reactive compound that inhibits DNA synthesis (Kalf, Post, and Snyder 1987).

Hemopoietic Toxicity - Aplastic anemia from benzene poisoning could arise from toxic damage to one or more of the components of the hemopoietic system: stem cells, transit cells (progenitor cells in various degrees of differentiation), and/or bone marrow stroma or microenvironment (Kalf, Post, and Snyder 1987).

Human Carcinogenicity - The weight-of-evidence classification for benzene is "A; human carcinogen." Several studies of increased incidence of nonlymphocytic leukemia from occupational exposure, increased incidence of neoplasia in rats and mice exposed by inhalation and gavage, and some supporting data form the basis for this classification (IRIS 1990). Aksoy et al. (1974); as cited in IRIS (1990), reported effects of benzene exposure among 28,500 Turkish workers employed in the shoe industry. Mean duration of employment was 9.7 years and mean age was 34.2 years. Peak exposure was reported to be 210 to 650 ppm. Twenty-six cases of leukemia and a total of 34 leukemias or preleukemias were observed, corresponding to an incidence of

13/100,000 (by comparison to 6/100,000 for the general population). A followup paper (Aksoy 1980), as cited in IRIS, reported eight additional cases of leukemia as well as evidence suggestive of increases in other malignancies (IRIS 1990). Several cohort mortality studies were cited in IRIS which showed significant increases in cases of leukemia associated with benzene exposures. Wong et al. (1983) as cited in IRIS, reported the mortality of male chemical workers who had been exposed to benzene for at least 6 months during the years 1946-1975. The study population of 4062 persons was drawn from seven chemical plants, and jobs were categorized as to peak exposure. Those with at least 3 day/wk exposure (3036 subjects) were further categorized on the basis of an 8-hour time weighted average. Dose-dependent increases were seen in leukemia and lymphatic and hematopoietic cancer. The incidence of leukemia was responsible for the majority of the increase (IRIS 1990). IRIS mentioned that numerous other epidemiologic and case studies have reported an increased incidence or a causal relationship between leukemia and exposure to benzene. Numerous investigators have found significant increases in chromosomal aberrations of bone marrow cells and peripheral lymphocytes from workers exposed to benzene (IARC 1982, as cited in IRIS 1990). Benzene also induced chromosomal aberrations in bone marrow cells from rabbits (Kissling and Speck 1973; as cited in IRIS 1990), mice (Meyne and Legator 1980; as cited in IRIS 1990), and rats (Anderson and Richardson 1979; as cited in IRIS 1990). Several investigators have reported positive results for benzene in mouse micronucleus assays (Meyne and Legator 1980; as cited in IRIS 1990). Benzene was not mutagenic in several bacterial and yeast systems, in the sex-linked recessive lethal mutation assay with Drosophila melanogaster, or in mouse lymphoma cell forward mutation assay (IRIS 1990).

Animal Carcinogenicity Data - Both gavage and inhalation exposure of rodents to benzene have resulted in development of neoplasia. Several studies were cited in IRIS regarding gavage doses with dose-related increases in the incidences of tumors. Slightly increased incidences of hematopoietic neoplasms were reported for male C57B1 mice exposed by inhalation to 300 ppm benzene 6 h/day, 5 day/wk for 488 days. There was no increase in tumor incidence in male AKR or CD-1 mice similarly exposed to 100 ppm or 300 ppm benzene, respectively (IRIS 1990).

Quantitative Estimate of Carcinogenic Risk From Inhalation Exposure -
The inhalation slope factor is 2.9E-2/mg/kg/day. The inhalation unit risk is
8.3E-6/μg/m^3. The extrapolation method used was the one-hit model using
pooled data. Table 2.2 shows air concentrations at specified risk levels
(IRIS 1990).

The unit risk estimate is the geometric mean of four point estimates
using pooled data from the Rinsky et al. (1981) and Ott et al. (1978) studies,
which was then adjusted for the results of the Wong et al. (1983) study; all
cited in IRIS (1990). The Rinsky data used were from an updated tape which
reports one more case of leukemia than was published in 1981. Equal weight
was given to cumulative dose and weighted cumulative dose exposure categories
as well as to relative and absolute risk model forms. The results of the Wong
et al. (1983) study were incorporated by assuming that the ratio of the
Rinsky-Ott-Wong studies to the Rinsky-Ott studies for the relative risk
cumulative dose model was the same as for other model-exposure category combinations and multiplying this ratio by the Rinsky-Ott geometric mean. The
age-specific U.S. death rates for 1978 (the most current year available) were
used for background leukemia and total death rates. It should be noted that a
recently published paper (Rinsky et al. 1987), reported yet another case of
leukemia from the study population. The unit risk should not be used if the
air concentration exceeds 100 μg/m^3, since above this concentration the slope
factor may differ from that stated (IRIS 1990).

Confidence of Inhalation Exposure Carcinogenicity - The pooled cohorts
were sufficiently large and were followed for an adequate time period. The
increases in leukemias were statistically significant and dose-related in one
of the studies. Wong et al. 1983 as cited in IRIS (1990), disagrees that
exposures reported in Rinsky et al. (1981) as cited in IRIS (1990), were
within the recommended standards. For the five leukemia deaths in persons
with 5 or more years exposure, the mean exposure levels (range 15 to 70 ppm)
exceeded the recommended standard (25 ppm) in 75% of the work locations
sampled. The risk estimate above based on reconsideration of the Rinsky et
al. (1981) and Ott et al. (1978) studies; as cited in IRIS (1990), is very
similar to that of 2.4E-2/ppm (cited in EPA 1980) based on Infante et al.

(1977a,b), Ott et al. (1978) and Aksoy et al. (1974); all cited in IRIS
(1990). It was felt by the authors of U.S. EPA (1985); as cited in IRIS
(1990), that the exposure assessment provided by Aksoy was too imprecise to
warrant inclusion in the current risk estimate. A total of 21 unit risk
estimates were prepared using six models and various combinations of the
epidemiologic data. These range over slightly more than one order of
magnitude. A geometric mean of these estimates is 2.7E-2/ppm. Regression
models give an estimate similar to the geometric mean (IRIS 1990).

Toxicant Interactions - Benzene metabolism and therefore benzene
toxicity is altered by simultaneous exposure to some other solvents (e.g.,
xylene, toluene) because these aromatic solvents are oxidized by many of the
same hepatic enzyme systems. Reported hematotoxic effects of benzene in
humans may be a synergistic result of simultaneous exposure to other solvents.
Since benzene metabolites rather than the parent compound are suspected of
inducing bone marrow toxicity, inhibition of benzene metabolism (hydroxyla-
tion) by toluene may result in increased toxic effects of the parent compound
instead of benzene metabolites (EPA 1984a).

2.2.2 n-Butylacetate

The contaminant n-butylacetate is a colorless liquid with a density of
0.88 and a flash point of 72°F. Its molecular formula is $C_6H_{12}O_2$, and its
molecular weight is 116.18. It has a vapor pressure of 15 mm at 25°C. It has
a hazard rating of 2 (Sax and Lewis 1989). The TLV on a TWA basis, for n-
butylacetate is 150 ppm. The TLV, on a short term exposure limit (STEL)
basis, is 200 ppm (ACGIH 1988).

2.2.2.1 Summary

Butylacetates are irritations and narcotic in high concentrations.
N-butylacetate has no notable systemic toxicity, but its vapor causes irritation of eyes and nose. The primary target organs for n-butylacetate are the
eye, skin, and respiratory system..

Persons with skin disease, kidney disease, chronic respiratory disease,
and liver disease may be at an increased risk from butylacetate exposure.

2.2.2.2 <u>Health Effects</u>

<u>General Toxicity</u> - n-butylacetate has no notable systemic toxicity, but its vapor causes irritation of eyes and nose. It may cause conjunctivitis. Concentrations of 1600 ppm have produced eye irritation in animals. Liquid application to rabbit eyes caused superficial injury graded 5 on a scale of 1 to 10. Workers exposed to greater than 200 ppm for 8 hours developed eye, nose, and throat irritations. n-butylacetate is normally first noticeable to humans at concentrations of 300 ppm in air and is objectionable at concentrations of 3300 ppm. Higher concentrations cause tearing and hyperemia of conjunctiva. Pulmonary edema has been reported in animals exposed to butyl acetates (HSDB 1990).

Butylacetates are narcotic in high concentrations. Concentrations of 1.4% caused death within 240 minutes in laboratory animals, and n-butylacetate has narcotic properties estimated at 1.7 times that of ethylacetate. Cerebral edema has occurred in laboratory animals (HSDB 1990).

Exposure of albino rabbits to 500 ppm for 20 days and to 1000 ppm for four days caused no corneal or conjunctival injury, as could be detected by slit-lamp biomicroscopy, and corneal sensation was not altered (HSDB 1990).

Slight central nervous system (CNS) depressant effects were noted in cats exposed for 6 hours at 6100 ppm of n-butylacetate. Slight irritation of eyes and salivation occurred at a concentration of 1600 ppm. The animals exposed 6 hours a day for 6 days at 3100 ppm concentration showed blood changes. Similar exposures at 4200 ppm resulted in slight irritation of the respiratory passages. Air nearly saturated with n-butylacetate (approximately 10,000 ppm) was fatal to six rats in an 8-hour period, but no deaths occurred, at this concentration, in 4 hours of exposure. Guinea pigs showed eye irritation at 3300 ppm concentrations of n-butylacetate. The guinea pigs became unconscious after 9 hours at a concentration of 7000 ppm and died following 4 hours of exposure at 14,000 ppm administered at a temperature of 30°C. In repeated inhalations of 3100 to 4200 ppm for 6 hours per day for 6 days, mice became habituated to the irritation but showed some fatigue and loss of weight. The blood picture of the exposed mice showed an increase in formed elements and hemoglobin. Mouse response to the inhalation of n-butylacetate at concentrations of 7400 ppm for 3 hours resulted in CNS depression and recovery (HSDB 1990).

Effects on isolated tissue was demonstrated by combining butylacetate with choline. In combination with choline, butylacetate exhibited a contractile effect on isolated guinea pig ileum. Release of acetylcholine by butylacetate was indicated, and when butylacetate combined with muscle acetylcholine receptor, the response to acetylcholine was inhibited (HSDB 1990).

Chicken eggs were studied for teratogenic effects. When injected with a dose of 45 mg, eggs did not hatch. When 27 mg per egg were injected, 45% hatched, and with 9 mg, 60% hatched. Kidney damage and corneal lesions were observed in the chick embryos (HSDB 1990).

Hepatotoxicity - The HSDB contained a reference to a study conducted by Franco et al. for which a detection of hepatotoxicity was reported from occupational exposure to n-butylacetate (HSDB 1990).

Embryotoxicity - As part of a critical analysis of exploratory methods used in testing chemical agents for embryotoxicity, teratogenicity and mutagenicity, the oncogenic reaction of laboratory mice showed butylacetate to by embryotoxic when administered to DBA or C57BL pregnant mice (HSDB 1990).

Mutagenicity - The mutagenicity of 43 industrial chemicals, including n-butylacetate, in Salmonella typhimurium and Escherichia coli was examined. The mutation test was performed in the absence and presence of rat microsomal activation. No mutagenic activity was observed with n-butylacetate (HSDB 1990).

Symptoms and General Effects - The following are symptoms for various levels of exposure to n-butylacetate: 1) CNS - headache, muscle weakness, giddiness, ataxia, confusion, delirium, coma, 2) gastrointestinal - nausea, vomiting, diarrhea (odor of the alcohol in excreta), 3) irritation of skin, eyes, throat from vapor or liquid, 4) death from respiratory failure, 5) disturbances of cardiac rhythm, 6) occasional complications - gastrointestinal hemorrhage, renal damage with glycosuria, liver damage, cardiac failure, and pulmonary edema (HSDB 1990).

Inhalation of paint thinner containing acetate esters, including butyl-acetate, caused drunkenness and hallucination in human subjects. Blood chemistry was performed on workers who were exposed to a variety of fat sol-vents, including butylacetate. Normocytic normochronic anemia occurred and was attributed to butylacetate and butylalcohol. Plasma bicarbonate was also lowered because of liberation of acetic acid (HSDB 1990).

Toxicity Ranges - 10,000 ppm of n-butylacetate for 8 hours was 100% fatal in test animals. A 4-hour exposure to the same level showed no fatali-ties. Brief exposures of humans to 3,300 ppm of n-butylacetate caused marked irritation to eyes and nose. Mild irritation was reported after a brief exposure to 200 to 300 ppm concentrations. Throat irritations in human sub-jects were noticed at 200 ppm concentrations, and the throat irritations became quite severe at 300 ppm (HSDB 1990).

Toxicant Interactions - Tests of butylacetate vapor on animal eyes have been complicated by an admixture of butylalcohol. It is uncertain whether damage to the corneal epithelium reported to occur at high concentrations was caused by the butylacetate or the butylalcohol. Likewise, a report showed several cases of vacuolar keratitis among workers exposed to a mixture of vapor of butylacetate and isobutylalcohol; however, it is uncertain which compound -- the butylacetate or the isobutylalcohol -- was responsible.

2.2.3 Carbon Tetrachloride

Carbon tetrachloride is a heavy colorless liquid with an ethereal odor. It has a density of 1.597 and no registered flash point. Its molecular formula is CCl_4, and its molecular weight is 153.81. Carbon tetrachloride has a vapor pressure of 100 mm at 23.0°C. It has a Hazard Rating of 3 (Sax and Lewis 1989). The TLV, on a TWA basis, for carbon tetrachloride is 5 ppm (ACGIH 1988). Common synonyms for carbon tetrachloride are acritet, benzinoform, carbona, carbon chloride, carbon tet, freon 10, halon 104, methane tetrachloride, tetrachloromethane, perchloromethane, tetrafinol, tetraform, and tetrasol.

2.2.3.1 Summary

Carbon tetrachloride is toxic by all routes of exposure and is manifested by both liver and renal damage, which may be delayed for up to 2 to 3
days following exposure. Liver injury is probably greater following ingestion
and can be permanent even from acute exposure. Neurological sequelae of
cerebellar dysfunction can occur.

Carbon tetrachloride vapor is a narcotic and causes severe damage to the
liver and kidneys. In humans, most fatalities have been the result of renal
injury with secondary cardia failure. Liver damage occurs more often after
ingestion of the liquid than after inhalation of the vapor. Human fatalities
from acute renal damage have occurred after exposure for 1/2 to 1 hour to
concentrations of 1,000 to 2,000 ppm. Cardiac arrhythmias have been reported
(HSDB 1990).

The weight-of-evidence classification for carbon tetrachloride is "B2;
probable human carcinogen." The basis for this classification is carcinogenicity in rats, mice and hamsters. There is inadequate human carcinogenicity data available. There have been three case reports of liver tumors
developing after carbon tetrachloride exposure. Several studies of workers
who may have used carbon tetrachloride have suggested that these workers may
have an excess risk of cancer.

As little as 3 to 5 ml of carbon tetrachloride ingested orally has
resulted in death. Toxicity via inhalation has been noted following a
30-minute exposure to 160 ppm. Systemic toxicity may follow dermal exposure.

Alcohol has been clearly shown to potentiate the toxicity of carbon
tetrachloride.

2.2.3.2 Health Effects

Noncarcinogenic Effects - The Reference Dose (RfD) is based on the
assumption that thresholds exist for certain toxic effects such as cellular
necrosis, but may not exist for other toxic effects such as carcinogenicity.
In general the RfD is an estimate (with uncertainty spanning possibly an order
of magnitude) of a daily exposure to the human population that is likely to be
without an appreciable risk of deleterious effects during a lifetime. Oral

RfDs are available for carbon tetrachloride. Its critical effect is liver lesions and the RfDo is 7E-4 mg/kg/day. There is a medium confidence assigned to the RfDo. The principal study was well conducted and good dose-response was observed in the liver, which is the target organ for CCl_4. Therefore, a high confidence level is assigned to the study. Four additional subchronic studies support the RfDo, but reproductive and teratology endpoints are not well investigated. Thus, the database rates a medium confidence level, with the result being a medium confidence level in the RfDo. No Reference Dose for chronic inhalation exposure (RfDi) is available at this time (IRIS 1990).

Exposure to CCl_4 at 10 or 11 ppm for 180 minutes produced no effect on liver or kidney function. However, exposure to 49 ppm for 70 minutes produced an effect on liver function shown by a reduction of serum iron 1 to 2 days later in 2 of 4 subjects (HSDB 1990).

Individuals who have experienced short-term overexposure may experience delayed effects, including damage to the heart, liver, and kidneys. Symptoms of liver damage may result in jaundice and dark urine.

Human Carcinogenicity - The weight-of-evidence classification for carbon tetrachloride is "B2; probable human carcinogen." The basis for this classification is carcinogenicity in rats, mice, and.hamsters. Human carcinogenicity data are inadequate. There have been three case reports of liver tumors developing after carbon tetrachloride exposure. Several studies of workers who may have used carbon tetrachloride have suggested that these workers may have an excess risk of cancer.

Animal Carcinogenicity - Carbon tetrachloride has proved carcinogenic to all species evaluated (rats, mice and hamsters), producing hepatocellular carcinomas in all of these species. Several studies were reported in IRIS confirming the carcinogenicity in these animals (IRIS 1990).

Carcinogenicity Supporting Data - Carbontetrachlorice was not mutagenic either for S. typhimurium or E. coli (McCann et al. 1975; Simmon et al. 1977; Uehleke et al. 1976; as cited in IRIS 1990). Carbon tetrachloride in low concentrations did not produce chromatic or chromosomal aberrations in an epithelial cell line derived from rat liver (Dean and Hodson-Walker 1979; as cited in IRIS 1990). In vivo unscheduled DNA synthesis assays have likewise been negative (Mirsalis and Butterworth 1980; Mirsalis et al. 1983; as cited in IRIS 1990). Carbon tetrachloride produced mitotic recombination and gene conversion is cerevisiae, but only at concentrations which reduced viability to 10% (Callen et al. 1980; as cited in IRIS 1990). Carbon tetrachloride may be metabolized to reactive intermediates capable of binding to cellular nucleophilic macromolecules. Negative responses in bacterial mutagenicity assays may have been due to inadequate metabolic activation in the test systems (IRIS 1990).

Quantitative Estimate of Carcinogenic Risk - The inhalation slope factor is 1.3E-1/mg/kg/day. The inhalation unit risk is 1.5E-5/μg/m^3. The extrapolation method used was the linearized multistage procedure using extra risk. Table 2.3 shows air concentrations at specified risk levels (IRIS 1990).

Inhalation risk was calculated assuming an air intake of 20 m^3/day and 40% absorption rate by humans (EPA 1984; as cited in IRIS 1990). This absorption coefficient was based on 30% inhalation in monkeys, and 30% and 57% to 65% inhalation in humans. A range of estimates of unit risk for inhalation exposures for the four studies cited above was determined, with 1.5E-5/μg/m^3 calculated as the geometric mean for the unit risk. The unit risk should not be used if the air concentration exceeds 7E+2 μg/m^3, since above this concentration the slope factor may differ from that stated (IRIS 1990).

Hepatotoxicity - There is an age difference in susceptibility to carbon tetrachloride-induced hepatotoxicity. At intraperitoneal doses of 1 ml/kg, the increase in serum aspartate aminotransferase and triglyceride accumulation in the neonatal (1- to 14-day-old) rat was equivalent to that observed in the

TABLE 2.3.   Carbon Tetrachloride Air Concentrations
at Specified Risk Levels

Risk Level	Concentration
E-4 (i in 10,000)	7E+0 μg/m^3
E-5 (1 in 100,000)	7E-1 μg/m^3
E-6 (1 in 1,000,000)	7E-2 μg/m^3

adult rat; however, there was much less macromolecular binding and lipid peroxidation in the young rats, whereas blood acetate levels were 3 to 5 times higher (HSDB 1990).

Response of specific forms of hepatic microsomal cytochrome p-450 to carbon tetrachloride was studied by immunohistochemical techniques to assess the localization and specificity of action for this hepatotoxin. Liver sections were taken from control, phenobarbital-pretreated, and 3-methyl-cholanthrene-pretreated male rats that had received acute (3-hour) treatment with carbon tetrachloride. Diminished fluorescence and loss of intracellular homogeneity fluorescence were found in the centrilobular cells of liver sections from carbon tetrachloride-challenged rats incubated with anti-p450b. In liver sections from 3-methylcholanthrene-pretreated, carbon tetrachloride challenged rates incubated with anti-p450c, fluorescence was diminished in periportal cells but did not show loss of intracellular homogenicity (HSDB 1990).

Rats were injected with carbon tetrachloride (0.2ml/kg body weight) twice weekly for 4 weeks. Carbon tetrachloride treatment caused a significant increase in hepatic lipid perodide levels and significant decrease in hepatic glutathione transferase activities. These results show that chronic carbon tetrachloride administration to rats leads to the stimulation of hepatic lipid peroxidation, which seems to be the consequence of impaired cellular defence by gluthione-related enzymes (HSDB 1990).

Teratogenicity - Carbon tetrachloride is not teratogenic to rats exposed orally, subcutaneously, or via inhalation (HSDB 1990).

Symptoms and General Effects - The following are symptoms for various levels of exposure to carbon tetrachloride: 1) prompt nausea, vomiting, and abdominal pain; after ingestion, hematemesis and diarrhea; 2) headache, dizziness, confusion, drowsiness, and occasionally convulsions; 3) visual disturbances, sometimes consisting of a concentric restriction of the color fields without central scotomata (toxic amblyopia); 4) rapid progression of central nervous depression with deepening coma and death from respiratory arrest or circulatory collapse; 5) occasionally sudden death due to ventricular fibrillation; 6) in massive exposures the above symptoms merge with those outlined below, but central nervous depression may subside without sequelae, or an essentially asymptomatic interval of a few days may precede hepatorenal decompensation, 7) kidney and/or liver injury, 8) oliguria, albuminuria, anuria, gradual weight gain, edema. Death may occur within 1 week in the absence of effective supportive treatment, 9) anorexia, jaundice, and right upper quadrant pain due to an enlarged and tender liver, 10) carpopedal spasm that was relieved by calcium gluconate appears to be a very rare reaction (HSDB 1990).

Toxicity Ranges - As little as 3 to 5 ml of carbon tetrachloride ingested orally has resulted in death. Toxicity via inhalation has been noted following a 30 minute exposure to 160 ppm. Systemic toxicity may follow dermal exposure.

Inhalation Absorption Factors - Pertinent studies of pulmonary absorption of carbon tetrachloride in humans were not located in the available literature. A few studies on pulmonary absorption in experimental animals were found. Nielsen and Larsen 1965, as cited in EPA 1984b, stated that carbon tetrachloride is "readily absorbed" through the lungs, but the species studied was not reported. Lehmann and Hasegawa (1910), as cited in EPA 1984b, showed that the rate of absorption decreased with duration of exposure. Von Oettingen et al. (1949, 1950), as cited in EPA 1984b, studied blood concentrations in dogs following exposure to 15 or 20 g/L in air. Peak blood concentrations of ≈35 or ≈38 mg/L were attained after ≈300 minutes of exposure to 15 or 20 g/L in air, respectively (EPA 1984b).

Toxicant Interaction - Alcohol has been clearly shown to potentiate the toxicity of carbon tetrachloride. Traiger and Plass (1971), as cited in EPA 1984b, investigated the potentiation of carbon tetrachloride toxicity by methanol, ethanol and isopropanol in rats. The activity of serum glutonic pyruvic transaminase (SGPT) was monitored to evaluate hepatotoxicity. All three alcohols tested potentiated the toxicity of carbon tetrachloride, with isopropanol being the most potent. Neither carbon tetrachloride nor the alcohols alone elevated SGPT levels (EPA 1984b).

2.2.4 Chloroform

Chloroform is a heavy colorless liquid with an ethereal odor. It has a density of 1.49845 and no registered flash point. Its molecular formula is $CHCl_3$, and its molecular weight is 119.37. It has a vapor pressure of 100 mm at 10.4°C. It has a Hazard Rating of 3 (Sax and Lewis 1989). The half-life of chloroform in air is 80 days (EPA 1984c). From 49% to 77% of the chloroform present in the inspired air is absorbed by the lungs (EPA 1984c). The TLV, on a TWA basis, for chloroform is 10 ppm (ACGIH 1988).

2.2.4.1 Summary

Chloroform exposure may occur by oral, inhalation, or dermal routes. Chloroform is an irritant, and central nervous system and cardiac depressant. Delayed renal and hepatic toxicity may also occur. Conjunctivitis and blepharospasm may occur from exposure to vapors of chloroform. Respiratory depression, chemical pneumonitis, and pulmonary edema may occur. Central nervous system depression, headache, and anorexia have been noted from significant exposures to chloroform. Central hepatic necrosis has occurred 10 to 48 hours postingestion, and fatty degeneration and hepatomegaly have been noted. Renal damage and frequent urination have been reported. Chloroform may be embryotoxic, and it is listed as a suspected carcinogen (HSDB 1990).

Chloroform is classified as moderately toxic. A probable oral lethal dose for humans is 0.5 to 5 g/kg (between 1 ounce and 1 pint) for a 150-lb. person. The mean lethal dose is probably near 1 fluid ounce (44 g) (Gosselin et al. 1976, as cited in IRIS 1990). Also, it is a central nervous system depressant and a gastrointestinal irritant (Challen et al. 1958, as cited in IRIS 1990). Chloroform has caused rapid death attributable to cardiac arrest (IRIS 1990).

2.2.4.2 Health Effects

Noncarcinogenic Effects - The RfD is based on the assumption that thresholds exist for certain toxic effects such as cellular necrosis, but may not exist for other toxic effects such as carcinogenicity. In general, the RfD is an estimate to the human population (including sensitive subgroups) that is likely to be without an appreciable risk of deleterious effects during a lifetime. The RfDo for chloroform is 1E-2/mg/kg/day, with the critical effect being fatty cyst formation in the liver. The confidence level assigned to this RfDo is medium. The critical study (Heywood et al. 1979, as cited in IRIS 1990), was of chronic duration, used a fairly large number of dogs and measured multiple endpoints; however, only two treatment doses were used and no NOEL was determined. Therefore, confidence in the study is rated medium. Confidence in the database is considered medium to low; several studies support the choice of a lowest-observed-adverse-effect level (LOAEL), but a no-observed-effect level (NOEL) was not found. Thus, confidence in the RfDo is also considered medium to low (IRIS 1990).

A 33-year-old male who habitually inhaled chloroform for 12 years has psychiatric and neurologic symptoms of depression, loss of appetite, hallucination, ataxia, and dysarthria. Other symptoms for habitual use are moodiness, mental and physical sluggishness, nausea, rheumatic pain, and delirium (HSDB 1990).

Worker exposure to concentrations of chloroform of over 112 mg/m^3 have been reported to result in depression, ataxia, flatulence, irritability, and liver and kidney damage. An increased incidence of cardiac arrhythmias has been demonstrated during surgery in patients anesthetized with chloroform as compared with other anesthetic agents at vapor concentrations of 22,500 ppm (HSDB 1990).

Human Carcinogenicity - The weight-of-evidence classification for chloroform is "B2; probable human carcinogen," based on increased incidence of several tumor types in rats and three strains of mice. The human carcinogenicity data are inadequate. There are no epidemiologic studies of chloroform itself. Chloroform and other trihalomethanes are formed from the interaction of chlorine with organic material found in water. Several ecological and case control studies of populations consuming chlorinated drinking water in which chloroform was the major chlorinated organic show small significant increases in the risk of rectal bladder or colon cancer on an intermittent basis. Many other suspected carcinogens were also present in these water supplies.

Animal Carcinogenicity - The animal carcinogenicity data are considered sufficient. Chloroform has been tested for carcinogenicity in eight strains of mice, two strains of rats and in beagle dogs.

In a gavage bioassay (NCI 1976, as cited in IRIS 1990), Osborne-Mendel rats and B6C3F1 mice were treated with chloroform in corn oil 5 times/wk for 78 weeks. Fifty male rats received 90 or 125 mg/kg/day; females initially were treated with 125 or 250 mg/kg/day for 22 weeks and 90 or 180 mg/kg/day at 18 weeks; females were dosed with 200 or 400, raised to 250 or 500 mg/kg/day. A significant increase in kidney epithelial tumors was observed in male rats, and increases in hepatocellular carcinomas were highly highly significant in mice of both sexes. Liver nodular hyperplasia was observed in low-dose male mice not developing hepatocellular carcinoma. Hepatomas have also developed in female strain A mice and NLC mice gavaged with chloroform (Eschenbrenner and Miller 1945; Rudali 1967; as cited in IRIS 1990).

Jorgenson et al. (1985); as cited in IRIS 1990, administered chloroform (pesticide quality and distilled) in drinking water to male Osborne-Mendel rats and female B6C3F1 mice at concentrations of 200, 400, 900, and 1800 mg/L for 104 weeks. These concentrations were reported by the author to correspond to 19, 38, 81, and 160 mg/kg/day for rats and 34, 65, 130, and 263 mg/kg/day for mice. A significant increase in renal tumors in rats was observed in the highest dose group. The increase was dose-related. The liver tumor incidence in female mice was not significantly increased. This study was specifically designed to measure the effects of low doses of chloroform (IRIS 1990).

Chloroform administered in toothpaste was not carcinogenic to male C57B1, CBA, CF-1 or female ICI mice or to beagle dogs. Male ICI mice administered 60 mg/kg/day were found to have an increased incidence of kidney epithelial tumors (Roe et al. 1979; Heywood et al. 1979; as cited in IRIS 1990). A pulmonary tumor bioassay in strain A/St mice was negative, as was one in which newborn C57X DBA2/F1 mice were treated subcutaneously on days 1 to 8 of life (Theiss et al. 1957; Roe et al. 1968; as cited in IRIS 1990).

Most tests for genotoxicity of chloroform have been negative. These negative findings include covalent binding to DNA, mutation in Salmonella, a Drosophila self-linked recessive, tests for DNA damage, a micronucleus test,

and transformation of BHK cells. By contrast, one study demonstrated binding
of radiolabeled chloroform to calf thymus DNA following metabolism by rat
liver microsome (DiRenzo 1982; as cited in IRIS 1990). Chloroform caused
mitotic recombination in Saccharomyces (Callen et al. 1980; as cited in IRIS
1990) and sister chromatic exchange in cultured human lymphocytes and in mouse
bone marrow cells exposed in vivo (Morimoto and Koizumi 1983; as cited in IRIS
1990).

The carcinogenicity of chloroform may be a function of its metabolism to
phosgene, which is known to cross-link DNA. A host-mediated assay using mice
indicated that chloroform was metabolized in vivo to a form mutagenic to
Salmonella strain TA1537. Likewise, urine extracts from chloroform-treated
mice were mutagenic (Agustin and Lim-Sylianco 1979; as cited in IRIS 1990).

Quantitative Estimate of Risk - The inhalation slope factor is
$8.1E-2/mg/kg/day$. The inhalation unit risk is $2.3E-5/\mu g/m^3$. The extrapo-
lation method used was the linearized multistage procedure, using extra risk.
Table 2.4 shows air concentrations at specified risk levels (IRIS 1990).

Dose-Response Data - Dose-response data for carcinogenicity, for inhalation exposure, is provided in IRIS 1990. The test animals were the B6C3F1
female mouse, administered with the oral and gavage routes. The tumor type
was hepatocellular carcinoma. Table 2.5 shows the dose-response data (IRIS
1990).

This inhalation quantitative risk estimate is based on oral data. Above
doses are TWA; at the end of the assay males weighed 35 g, and females 28 g.
Exposure vehicle control animals were run concurrently and housed with test

TABLE 2.4. Chloroform Air Concentrations at
Specified Risk Levels

Risk Level	Concentration
E-4 (1 in 10,000)	$4 \ \mu g/m^3$
E-5 (1 in 100,000)	$4E-1 \ \mu g/m^3$
E-6 (1 in 1,000,000)	$4E-2 \ \mu g/m^3$

TABLE 2.5.   Chloroform Inhalation Exposure
Dose-Response Data

Dose		
Administered (mg/kg/day)	Human Equivalent (mg/kg/day)	Tumor Incidence
Female		
0	0	0/20
238	9.9	36/45
477	19.9	39/41
Male		
0	0	1/18
138	6.2	18/50
277	12.5	44/45

animals. All treated animals experienced decreased body weight gain.
Survival was reduced in high-dose males and in all treated females (IRIS
1990).

Experimental data for this compound support complete absorption of
orally administered chloroform under conditions of this assay. There are not
apparent species differences in this regard. Extrapolation of metabolism-
dependent carcinogenic responses from mice to humans on the basis of body
surface area is supported by experimental data. The slope factor is the
geometric mean calculated from male (3.3E-2) and female (2.0E-1) data. The
unit risk should not be used if the air concentration exceeds 400 $\mu g/m^3$, since
above this concentration the slope factor may differ from that stated. Slope
factors derived from male rat kidney tumor data (2.4E-2) (NCI 1976; as cited
in IRIS 1990) and studies by Roe et al. (1979); as cited in IRIS 1990,
(1.0E-1) are generally supportive of the risk estimate (IRIS 1990).

Teratogenicity - Teratogenic effects (acaudia, imperforate anus,
decreased crown-rump length, missing ribs, and delayed skeletal ossification)
were seen in Sprague-Dawley rats (Schwetz et al. 1974; as cited in EPA 1984c)
that inhaled chloroform for 7 h/day on days 6 to 15 of gestation at dose
levels of 30, 100 and 300 ppm. When CF/1 mice (Murray et al. 1979; as cited

in EPA 1984c) were exposed to 100 ppm chloroform for 7 h/day on days 6 to 15 of gestation, there was a significantly increased incidence of cleft palate. When pregnant mice and rats were exposed to 100 ppm chloroform, their food consumption and body weight decreased, but their relative liver weight increased (EPA 1982, as cited in EPA 1984c). Ingestion of chloroform caused fetotoxicity but not teratogenicity, and only at levels that also produced severe maternal toxicity (Thompson et al. 1974, as cited in EPA 1984c).

Toxicity Ranges - As little as 10 ml in an acute ingestion of chloroform may result in central nervous system depression and death (HSDB 1990).

Symptoms - Signs of chloroform poisoning in humans include a character- istic sweetish odor on the breath, dilated pupils, cold and clammy skin, initial excitation alternating with apathy, loss of sensation, abolition of motor functions, prostration, unconsciousness, and eventual death.

Toxicant Interactions - The substrates that potentiate the toxic effects of chloroform are methyl n-butyl ketone, alcohol, carbon tetrachloride, chlor- decone, DDT and phenobarbital (EPA 1984c). Methyl n-butyl ketone increases the toxicity of chloroform by lowering glutathione levels and by increasing the levels of hepatic cytochrome p-450, which in turn, increases the metabol- ism of chloroform to phosgene (Branchflower and Pohl 1981; as cited in EPA 1984c). Haris et al. (1982; as cited in EPA 1984c) reported that carbon tetrachloride potentiated the toxic effects of chloroform, because of increased phosgene formation and the initiation of lipid peroxidation. The mechanism of interaction for alcohol, chlordecone, DDT and phenobarbital was not discussed. Von Oettingen (1964; as cited in EPA 1984c) reported that high-fat/low-protein diets potentiated hepatotoxic effects of chloroform in animals (EPA 1984c).

2.2.5 Decane

Decane is a liquid with a density of 0.730 and a flash point of 115°F. Its molecular formula is $C_{10}H_{22}$, and its molecular weight is 142.29. It has a vapor pressure of 1 mm at 16.5°C. It has a Hazard Rating of 3 (Sax and Lewis 1989).

2.2.5.1 Summary

The most serious toxic effect following ingestion of decane is aspiration pneumonitis. Aspiration of hydrocarbons may also result in transient central nervous system depression or excitement. Secondary effects may include hypoxia, infection, pneumatocele formation, and chronic lung dysfunction. Inhalation may result in euphoria, cardiac dysrhythmias, respiratory arrest, and central nervous system toxicity.

2.2.5.2 Health Effects

The amount of information available regarding toxicological effects of decane is very limited. No specific studies involving the toxicology of decane were found.

Symptoms and General Effects - Coughing, choking, tachypnea, dyspnea, cyanosis, roles, hemoptysis, and pulmonary edema may occur following ingestion and aspiration. Transient central nervous system excitation followed by depression may occur, especially after inhalation. Liver injury, as manifested by elevated transaminases, may occur following ingestion, and renal injury, as manifested by tubular damage, may occur following ingestion (HSDB 1990).

Decane was oxidized by microsome from livers of mouse, rat, rabbit, beef, pigeon and chick embryo. Decanol, decanoic acid and decamethylene glycol were major metabolites of oxidation of decane (HSDB 1990).

Tumorigenic Data - The tumorigenic data provided in RTECS was very limited and sketchy. Mouse skin study with a low dose (25 gm/kg per 52 weeks) yielded that decane was an equivocal tumorigenic agent by Registry of Toxic Effects of Chemical Substances (RTECS) criteria. Tumors appeared at the site of application (RTECS 1987).

Toxicity Data - The toxicity data provided in RTECS were very limited and sketchy. Mouse inhalation studies yielded an LC50 of 72,300 mg/m^3/2 hr (RTECS 1987).

Toxicity Range - Less than 1 ml of some hydrocarbons, when directly aspirated into the lungs in animals, has produced severe pneumonitis (HSDB 1990).

Toxicant Interactions - Series of tobacco smoke components and related compounds were tested for carcinogenic activity on mouse skin. The compound was applied to mouse skin three times per week with low doses (5 μg/application) of benzo(a)pyrene [B(a)P]. Decane enhanced carcinogenicity of B(a)P (Van Douren et al. 1976, as cited in HSDB 1990).

2.2.6 1,2-Dichlorobenzene

1,2-Dichlorobenzene is a clear liquid with a density of 1.307 and a flash point of 151°F. Its molecular formula is $C_6H_4Cl_2$ and its molecular weight is 147.00. It has a Hazard Rating of 3 (Sax and Lewis 1989). The TLV, on a TWA basis, is 50 ppm. This TLV represents a ceiling limit. Some of the common synonyms for 1,2-dichlorobenzene are o-dichlorobenzene, chloroben, DCB, o-dichlor benzol, Dilantin DB, Dilatin DB, dizene, Dowtherm E, orthodichloro-benzene, orthodichlorobenzol, Termitkil, and ODB.

2.2.6.1 Summary

The RfDo is 9E-2 mg/kg/day, with no adverse critical effects observed (IRIS 1990). Inhalation of up to 400 ppm of o-dichlorobenzene was neither teratogenic nor fetotoxic in rats and neither o-dichlorobenzene or p-dichlorobenzene was teratogenic nor fetotoxic in rabbits at exposure levels up to 400 or 800 ppm, respectively (HSDB 1990).

A 55 year-old female received a non-occupational chronic repeated inhalation exposure to vapors from use of solution to clean clothes. The concentration was estimated at 1 to 2 L/yr. The result was acute myeloblastic leukemia (HSDB 1990).

2.2.6.2 Health Effects

Noncarcinogenic Effects - The RfD is based on the assumption that thresholds exist for certain toxic effects such as cellular necrosis, but may not exist for other toxic effects such as carcinogenicity. In general, the RfD is an estimate (with uncertainty spanning perhaps an order of magnitude) of a daily exposure to the human population (including sensitive subgroups) that is likely to be without an appreciable risk of deleterious effects during a lifetime. The RfDo is 9E-2 mg/kg/day, with no adverse critical effects observed (IRIS 1990).

1,2-dichlorobenzene in corn oil was given by gavage to F344/N rats and B6C3F1 mice (50 males and 50 females/group) at doses of 0, 60, or 120 mg/kg/ day, 5 day/wk for 103 weeks (NTP 1985, as cited in IRIS 1990). Survival of high-dose male rats was decreased compared with controls (19/50 versus 42/50, $P \geq 0.001$), but the difference appears largely because of deaths from gavage error (4 controls versus 20 high-dose). Although an increase ($P \geq 0.05$) in renal tubular regeneration in high-dose male mice was observed (control, 8/48; low-dose, 12/50; high-dose, 17/49), there was no other evidence of treatment-related renal lesions in either species. Further, control incidence of this lesion in male mice is below those of three similar control groups on study at the testing facility during approximately the same period (31/50, 15/50, 24/50). There was no other evidence of treatment-related effects in this study. Because the above differences between control and high-dose animals are questionable, a NOAEL of 120 mg/kg/day is concluded (IRIS 1990).

1,2-Dichlorobenzene in corn oil was given orally by gavage to F344/N rats and B6C3F1 mice (10 males and 10 females/group) at doses of 0, 30, 60, 125, 250, or 500 mg/kg/day, 5 day/wk for 13 weeks (NTP 1985; as cited in IRIS 1990). Liver necrosis was found in mice and rats given 250 mg/kg/day. Deaths, degeneration and necrosis in liver, lymphocyte depletion in spleen and thymus, renal tubular degeneration (male rats only), and slight decreases in hemoglobin, hematocrit and red blood cell counts (rats only) were induced with 500 mg/kg/day (IRIS 1990).

Necrosis (focal or individual hepatocyte) was observed in the livers of one male and three female rats given 125 mg/kg/day (NTP 1985, as cited in IRIS 1990). Increases ($P \geq 0.05$) in serum cholesterol were observed at all doses except 60 mg/kg/day in male rats and 30 and 60 in liver; increases in body weight ratios were observed in male and female rats, serum protein at all doses in female rats, and treatment-related liver effects at doses >125 mg/kg/day. However, no evidence of treatment-related liver pathology was observed in rats and mice given 60 or 120 mg/kg/day, 5 day/wk in the 2-year National Toxicology Program (NTP 1985, as cited in Iris 1990), carcinogenicity bioassay and no increase (P≥0.05) in serum enzymes (SGPT, GGPT, alkaline phosphatase) are grounds for considering 125 mg/kg/day as a NOAEL in the 13-week study (IRIS 1990).

In rats dosed by gavage with 1,2-dichlorobenzene at 18.8, 188, or 376 mg/kg/day, 5 day/wk for 192 days, the NOAEL was 18.8 mg/kg/day, liver and kidney weights were increased at 188 mg/kg/day, and liver pathology and increased spleen weight were observed with 375 mg/kg/day (Hollingsworth et al. 1958; as cited in IRIS 1990).

Rats, guinea pigs, mice, rats, and monkeys were exposed by inhalation to 1,2-dichlorobenzene at levels of 49 or 93 ppm, 7 h/day, 5 day/wk for 6 to 7 months. At 93 ppm, body weight gain in rats and spleen weight in guinea pigs were reduced (P≥0.05) (Hollingsworth et al. 1958, as cited in IRIS 1990). Estimated daily doses with 49 ppm exposure are 387 mg/kg (mouse), 19.3 mg.kg (rat), 14.4 mg/kg (guinea pig), 15.9 mg/kg (rabbit), and 20.3 mg/kg (monkey) (IRRIS 1990).

Pregnant F344/N rats and New Zealand rabbits were exposed by inhalation to 0, 100, 200, or 400 ppm to 1,2-dichlorobenzene for 6 hours daily on days 6 through 15 (rats) or 6 through 18 (rabbits) of gestation (Hayes 1985, as cited in IRIS 1990). Body weight gain was lower (P≥0.05) in rats at all doses and in rabbits, during the first three days of exposure, at 400 ppm. Liver weight (absolute and relative to body weight) was increased in rats at 400 ppm. No developmental toxicity was evident at any dose. Estimated daily doses with 100 ppm exposure are 40 mg/kg (rat) and 32 mg/kg (rabbit) (IRIS 1990).

No RfDi is available at this time.

Seventeen chemicals (solvents, insecticides and intermediates used in the production of textiles and resins) were tested in a short-term in vitro system with human lymphocytes to determine their action. The parameters studied were tritiated thymidine uptake and cell viability in cultures grown

with or without a rat liver metabolizing system (S-9 mix). 1,3-dichloro-
benzene, 1,2-dichlorobenzene, hexane, 1,2-diiodoethane, 1,4-dichlorobenzene,
tetrachloroethylene, 2,3-dibromopropanol, chloromethyl methyl ether, 1,2- and
1,3-dibromopropane, in order, exerted the more toxic effects. The chemicals
were non-toxic in the presence of the metabolizing system with exception of
1,2- and 1,3-dichlorobenzene which maintained, to some degree, their toxicity
even in the presence of the S-9 mix (HSDB 1990).

Leukemia - A 55 year-old female received a non-occupational chronic
repeated inhalation exposure to vapors from use of solution to clean clothes.
The concentration was estimated at 1 to 2 L/yr. The result was acute
myeloblastic leukemia (HSDB 1990).

Teratogenic Potential - o-dichlorobenzene (1,2-dichlorobenzene) and
p-dichlorobenzene (1,4-dichlorobenzene) were evaluated for teratogenic
potential in rats (o-dichlorobenzene only) and rabbits. Groups of bred rats
and inseminated rabbits were exposed to 0, 100, 200, or 400 ppm of o-dichlorobenzene, while groups of inseminated rabbits were exposed to 0, 100, 300, or
800 ppm p-dichlorobenzene. The animals were exposed for 6 h/day on days
6 through 15 (rats) or days 6 through 18 (rabbits) of gestation. Maternal
toxicity, as evidenced by a significant decrease in body weight gain, was
observed in all groups of o-dichlorobenzene exposed rats and liver weight was
significantly increased in the 400 ppm o-dichlorobenzene exposed group.
Slight maternal toxicity was observed in groups of rabbits exposed to 400 ppm
o-dichlorobenzene or 800 ppm p-dichlorobenzene as indicated by significantly
decreased body weight gain during the first three days of exposure. Inhalation of up to 400 ppm of o-dichlorobenzene was neither teratogenic nor
fetotoxic in rats. Neither o-dichlorobenzene nor p-dichlorobenzene was
teratogenic or fetotoxic in rabbits at exposure levels up to 400 or 800 ppm,
respectively (HSDB 1990).

Toxicity Ranges - Toxic doses vary enormously with route and rate of
exposure (HSDB 1990).

2.2.7 Dichloromethane

Dichloromethane is a colorless volatile liquid with a density of 1.326. Its molecular formula is CH_2Cl_2, and its molecular weight is 84.93. Dichloromethane has a vapor pressure of 380 mm at 22°C. It has a Hazard Rating of 3 (Sax and Lewis 1989). The TLV, on a TWA basis, for dichloromethane is 50 ppm (ACGIH 1988). Common synonyms for dichloromethane are methylene chloride, methane dichloride, DCM, 1,1-dichloromethane, Freon 30, methylene bichloride, and methylene dichloride.

2.2.7.1 Summary

The RfDo for dichloromethane is 6E-2mg/kg/day, with liver toxicity as the critical effect. The weight-of-evidence classification for dichloromethane is "B2; probable human carcinogen," based on inadequate data in humans and increased cancer incidence in rats and mice (IRIS 1990). The inhalation slope factor is 1.4E-2/mg/kg/day. The inhalation unit risk is 4.1E-6/μg/m^3. The extrapolation method used was linearized multistage procedure, with extra risk considered.

There is concern about consumer exposure to dichloromethane from products because the dichloromethane retained in inhalation is metabolized to carbon monoxide, leading to elevated levels of carboxyhemoglobin. At relatively high exposures, the resulting anoxic stress placed on individuals may pose a health hazard. In addition, a recent NTP inhalation bioassay found clear evidence of carcinogenicity in female rats and in both sexes of mice as well as "some evidence of carcinogenicity" in male rats (Girman and Hodgson 1986).

2.2.7.2 Health Effects

Noncarcinogenic Effects - The RfD is based on the assumption that thresholds exist for certain toxic effects such as cellular necrosis, but may not exist for other toxic effects such as carcinogenicity. In general, the RfD is an estimate of a daily exposure to the human population (including sensitive subgroups) that is likely to be without an appreciable risk of deleterious effects during a lifetime. The RfDo for dichloromethane is 6E-2mg/kg/day, with liver toxicity as the critical effect.

The chosen study supporting the RfDo was a 24-month chronic toxicity and oncogenicity study of methylene chloride in rats (National Coffee Association 1982; as cited in IRIS 1990). This study appears to have been very well conducted, with 85 rats per sex at each of four nominal dose groups (i.e., 5, 50, 125, and 250 mg/kg/day) for 2 years. A high-dose recovery group of 25 rats per sex, as well as two control groups of 85 and 50 rats per sex, was also tested. Many effects were monitored. Treatment related histological alterations of the liver were evident at nominal doses of 50 mg/kg/day or higher. The low nominal dose of 5 mg/kg/day was a NOAEL (IRIS 1990).

The supporting database is limited for the RfDo. A NOAEL of 87 mg/m^3 was reported in one inhalation study (Haun et al. 1972, as cited in IRIS 1990). The equivalent oral dose is about 28 mg/kg bw/day (i.e., 87 mg/m^3 x 0.05 x 0.223 m^3/day/0.35 kg; these exposure values are for rats (IRIS 1990).

The confidence rating for the dichloromethane RfDo is medium. The study supporting the RfDo is given a high confidence rating because a large number of animals of both sexes were tested in four dose groups, with a large number of controls. Many effects were monitored and a dose-related increase in severity was observed. The database is rated medium to low because only a few studies support the NOAEL. Thus, medium confidence in the RfDo follows.

The RfDi is not available at this time.

Human Carcinogenicity - The weight-of-evidence classification for dichloromethane is "B2; probable human carcinogen," based on inadequate data in humans and increased cancer incidence in rats and mice (IRIS 1990).

Human carcinogenicity data are inadequate. Neither of two studies of chemical factory workers showed an excess of cancers (Friedlander et al. 1978, 1985; Ott et al. 1983, as cited in IRIS 1990). In the former study, exposures were low, but the data provided some suggestion of an increased incidence of pancreatic tumors. The latter report was designed to examine cardiovascular effects, and the study period was too short to allow for latency of site-specific cancers (IRIS 1990).

Animal Carcinogenicity Data - The animal carcinogenicity data are considered sufficient. In a 2-year study (National Coffee Association 1982,

1983, as cited in IRIS 1990), F344 rats received 0, 5, 50, 125, or 250 mg dichloromethane/kg/day in drinking water. B6C3F1 mice consumed 0, 60, 125, 185, or 250 mg/kg/day in water. Female rats responded with increased incidence of neoplastic nodules or hepatocellular carcinomas, which was significant by comparison to matched but not to historical controls. Male rats did not show an increased incidence of liver tumors. Male mice had elevated incidences of combined neoplastic modules and hepatocellular carcinomas, but female mice did not. This increase was not statistically significant or dose-related. In a study by the NTP (1982), as cited in IRIS (1990), gavage study of rats and mice has not been published because of data discrepancies (IRIS 1990).

Inhalation exposure of male and female Syrian hamsters to 0, 500, 1500, or 3500 ppm dichloromethane for 6 h/day, 5 day/wk for 2 years did not produce neoplasia. Female Sprague-Dawley rats exposed under the same conditions experienced reduced survival at the highest dose. Increased incidences of mammary tumors were noted in both males and females. Male rats also developed salivary gland sarcomas (Burek et al. 1984, as cited in IRIS 1990). There is a question as to whether these doses were at or near the maximum tolerated dose (MTD). In a subsequent study (Burek et al. 1984, as cited in IRIS 1990) male and female rats were exposed to 0, 50, 200 or 500 ppm dichloromethane. No salivary tumors were observed, but the highest dose resulted in mammary tumors (IRIS 1990).

Groups of 50 each male and female F344/N rats and B6C3F1 mice were exposed to dichloromethane 6 h/day, 5 day/wk for 2 years. Exposure concentrations were 0, 1000, 2000, or 4000 ppm for rats and 0, 2000, or 4000 ppm for mice. Survival of male rats was low, but apparently not treatment related; survival was decreased in a treatment-related fashion for male and female mice and female rats. Mammary adenomas and fibroadenomas were increased in male and female rats as were mononuclear cell leukemias in female rats. Among treated mice of both sexes, there were increased incidences of hepatocellular adenomas and carcinomas and highly significant increases in alveolar/bronchiolar adenomas and carcinomas (NTP 1986, as cited in IRIS 1990).

Two inhalation assays using dogs, rabbits, guinea pigs, and rats were negative, but were not carried out for the lifetime of the animals (Heppel et al. 1944; MacEwen et al. 1972, as cited in IRIS 1990). Theiss et al. (1977), as cited in IRIS 1990, injected strain A male mice intraperitoneally with 0, 160, 400, or 800 mg/kg for 16-17 times. Pulmonary adenomas were found, but survival of animals was poor (IRIS 1990).

As supporting data for carcinogenicity, dichloromethane is mutagenic for Salmonella typhimurium with or without added hepatic enzymes (Green 1983, as cited in IRIS 1990) and produced mitotic recombination in yeast (Callen et al. 1980, as cited in IRIS 1990). Results in cultured mammalian cells have generally been negative, but dichloromethane has been shown to transform rat embryo cells and to enhance viral transformation of Syrian hamster embryo cells (Price et al. 1978; Hatch et al. 1983, as cited in IRIS 1990).

Quantitative Estimate of Carcinogenic Risk - The oral slope factor is 7.5E-3/mg/kg/day. The extrapolation method used was linearized multistage procedure, with extra risk considered. The inhalation unit risk is 4.7E-7/μg/m^3. An inhalation slope factor is not available because pharma-cokinetic models were used to estimate the unit risk value. Table 2.6 shows air concentrations at specified risk levels (IRIS 1990).

Dose-Response Data - Inhalation dose-response data for carcinogenicity is provided in IRIS (1990). The test animal was the female B6C3F1 mouse; scientists looked at combined carcinomas and adenomas of the lung or liver. Table 2.7 shows the dose-response data (IRIS 1990).

Dose conversions used the mouse assay midpoint weight of 0.032 kg and estimated inhalation rate of 0.0407 m^3/day. To obtain estimates of unit risk

TABLE 2.6. Dichloromethane Air Concentrations at Specified Risk Levels

Risk Level	Concentration
E-4 (1 in 10,000)	2E+2 μg/m^3
E-5 (1 in 100,000)	2E+1 μg/m^3
E-6 (1 in 1,000,000)	2 μg/m^3

TABLE 2.7. Dichloromethane Inhalation Exposure
Dose-Response Data

Dose		
Administered (mg/kg/day)	Human Equivalent (mg/kg/day)	Tumor Incidence
Female		
0	0	5/50
1582	122	36/48
3164	244	46/57

for humans, an inhalation rate of 20 m^3/day was assumed. Dichloromethane was considered to be a well-absorbed vapor at low doses. A revision of the cancer risk assessment was released in September 1990. This revision contains a new inhalation unit risk based on the incorporation of information on pharmacokinetics and metabolism. This unit risk should not be used if the air concentration exceeds 2E+4 $\mu g/m^3$, because above this concentration, the slope factor may differ from that stated (IRIS 1990).

Adequate numbers of animals were observed and tumor incidences significantly increased in a dose-dependent fashion. Analysis, excluding animals that died before observation of the first tumors, produced similar risk estimates as did time-to-tumor analysis. However, significant uncertainty still exists in the inhalation unit risk because of uncertainty in the pharmacokinetic model for dichloromethane.

2.2.8 Dodecane

Dodecane has a molecular formula of $C_{12}H_{26}$, and a molecular weight of 170.38. It has a Hazard Rating of 3 (Sax and Lewis 1989). Some common synonyms for dodecane are adakane 12, bihexyl, dihexyl, n-dodecane, and duodecane.

2.2.8.1 Summary

Dodecane, which is not highly toxic, is a possible potentiator of skin tumorigenesis by B(a)P, decreasing the effective threshold dose by a factor of 10 (Patty 1981, as cited in HSDB 1990).

The RTECS database lists a mouse skin Toxic Dose Low (TDLo) of 11 g/kg/22 weeks.

2.2.8.2 Health Effects

The amount of information available regarding toxicological effects of dodecane is very limited. The HSDB database lists dodecane, but does not provide any information regarding human toxicity. The IRIS database does not list dodecane.

Tumorigenic Data - The RTECS database lists a mouse skin TDLo of 11 g/kg/22 weeks. RTECS has dodecane listed as a "equivocal tumorigenic agent by RTECS criteria" (RTECS 1987).

Non-Human Toxicity - An isolated perfused rabbit lung preparation was used to study the influence of pretreatment with a cocarcinogen, n-dodecane, on the metabolism of B(a)P. B(a)P was administered intratracheally to the isolated perfused lung following biweekly inhalation exposure of the rabbit to n-dodecane. N-dodecane appears to be as good an enzyme inducer as B(a)P in stimulating metabolism of B(a)P in isolated perfused rabbit lung (HSDB 1990).

A ratio of potencies of dodecane, a lymphocyte comitogen, has previously been found for promotion of mouse epidermal tumorigenesis. Maximum comitogenic activity was found when alkane and lectin were added to cultures simultaneously, the effect decreasing sharply when alkane addition was delayed relative to lectin. The existence of a cellular receptor with specificity for hydrophobic functions with chain length lying within a restricted range was suggested. Binding to this a receptor also appears to be a common early event in tumor promotion, comitogenicity, and comutagenicity, by agents of several chemical types (Baxter et al. 1981, as cited in HSDB 1990).

In an isolated rabbit heart, mitochondria incubation (at 15-38 degrees) with n-dodecane at 10-160 μg/mg mitochondrial protein resulted in uncoupling of oxidative phosphorylation, but only at 30-38 degrees. A slight inhibition of hydronicotinamide adenine dinucleotide (NADH) oxidase was also noted. N-dodecane exerted a biphasic action activation followed by inhibition of succinate oxidase (Borgatti et al. 1981, as cited in HSDB 1990).

Toxicant Interactions - Dodecane and phenyldodecane applied topically to progeny of rats treated with B(a)P, Chrysene or benzo(b)triphenylene on the 17th day of gestation produced tumors in offspring (Tanaka 1978, as cited in HSDB 1990).

Dodecane was shown to be a potentiator of some carcinogens above an apparent 0.02% threshold dose when, for example, B(a)P dissolved in decalin was applied to the mouse skin. Dodecane, which is not highly toxic, is a possible potentiator of skin tumorigenesis by B(a)P, decreasing the effective threshold dose by a factor of 10 (Patty 1981, as cited in HSDB 1990).

2.2.9 2-Ethoxyethylacetate

2-ethoxyethylacetate (cellosolve acetate) is a colorless liquid with a mild, pleasant ester-like odor. It has a density of 0.9748, and a flash point of 117°F. Its molecular formula is $C_6H_{12}O_3$ and its molecular weight is 132.18. It has a vapor pressure of 1.2 mm at 20°C. It has a Hazard Rating of 2 (Sax and Lewis 1989). The TLV, on a TWA basis, is 5 ppm (ACGIH 1988). There are several synonyms for 2-ethoxyethylacetate; some of the more common ones are cellosolve acetate, 2-ethoxyethanol acetate, ethylene glycol, ethyl ether acetate, ethylglycol acetate, and oytol acetate.

2.2.9.1 Summary

The amount of information regarding toxicological effects of 2-ethoxyethylacetate is very limited. No studies of human toxicology were found.

2.2.9.2 Health Effect

The amount of information regarding toxicological effects of 2-ethoxyethylacetate is very limited. No studies of human toxicology were found.

2-ethoxyethylacetate is moderately toxic by ingestion and intraperitoneal routes. It is mildly toxic by skin contact, inhalation and subcutaneous routes. It is a skin and eye irritant and an experimental teratogen (Sax and Lewis 1989).

General Toxicity - 2-ethoxyethylacetate is moderately toxic by ingestion and intraperitoneal routes. It is mildly toxic by skin contact, inhalation and subcutaneous routes. It is a skin and eye irritant and an experimental teratogen. Other experimental reproductive effects have been observed (Sax and Lewis 1989).

2.2.10 Ethylbenzene

Ethylbenzene is a colorless liquid with an aromatic odor. It has a density of 0.8669, and a flash point of 59°F. Its molecular formula is C_8H_{10} and its molecular weight is 106.18. It has a vapor pressure of 10 mm at 25.9°C (Sax and Lewis 1989). The TLV, on a TWA basis, for ethylbenzene is 100 ppm. The TLV, on a short-term exposure basis (STEL) basis, is 125 ppm (ACGIH 1988). Some of the common synonyms for ethylbenzene are aethylbenzol, EP, ethylbonzol, and phenylethane.

2.2.10.1 Summary

The RfDo for ethylbenzene is 1E-1mg/kg/day, with the critical effect of liver and kidney toxicity. The LOAEL of 408 mg/kg/day is associated with histopathologic changes in liver and kidney (IRIS 1990).

Prolonged exposure to vapors of ethylbenzene may result in functional disorders, increase in deep reflexes, irritation of the upper respiratory tract, hematological disorders (leukopenia and lymphocytosis, in particular), and hepatobiliary complaints (HSDB 1990). Aspiration of even a small amount of ethylbenzene may cause severe injury because its low viscosity and surface tension cause it to spread over a large surface of pulmonary tissue (HSDB 1990).

Ethylebenzene produces an irritant effect from chronic inhalation at 100 ppm/8 h (Patty 1981, as cited in HSDB).

It has been shown that the concentration of 1 mg/L and even 0.1 mg/L of ethylbenzene may be dangerous and may produce functional and organic disturbances (i.e., nervous system disorders, toxic hepatitis and upper respiratory tract complaints). Concentrations as low as 0.01 mg/L may lead to inflammation of upper respiratory tract mucosa (HSDB 1990).

Ethylbenzene vapor has a transient irritant effect on human eyes at 200 ppm in air. At 1000 ppm on the first exposure, it is very irritating and caused tearing, but tolerance rapidly develops. At 2000 ppm, eye irritation and lacrimation are immediate and severe. Concentrations of 5000 ppm cause intolerable irritation of the eyes and nose (HSDB 1990).

The weight-of-evidence classification for ethylbenzene is "D; not classifiable as to human carcinogenicity."

2.2.10.2 Health Effects

Noncarcinogenic Effects - The RfD is based on the assumption that thresholds exist for certain toxic effects such as cellular necrosis, but may not exist for other toxic effects such as carcinogenicity. In general, the RfD is an estimate of a daily exposure to human population (including sensitive subgroups) that is likely to be without an appreciable risk of deleterious effects during a lifetime. The RfDo for ethylbenzene is 1E-1 mg/kg/day, with the critical effect of liver and kidney toxicity. The criteria considered in judging the toxic effects on the test animals were growth, mortality, appearance and behavior, hematologic findings, terminal concentration of urea nitrogen in the blood, final average organ and body weights, histopathologic findings, and bone marrow counts. The LOAEL of 408 mg/kg/day is associated with histopathologic changes in liver and kidney (IRIS 1990).

The confidence in the RfDo is low (IRIS 1990). Confidence in the chosen study is low because rats of only one sex were tested and the experiment was not of chronic duration. Confidence in the supporting database is low because other oral toxicity data were not found.

The RfDi is not available at this time.

Erythema and inflammation of skin may result from contact of the skin with liquid (Sax and Lewis 1989).

Prolonged exposure to vapors of ethylbenzene may result in functional disorders, increase in deep reflexes, irritation of the upper respiratory tract, hematological disorders (leukopenia and lymphocytosis, in particular), and hepatobiliary complaints (HSDB 1990).

Aspiration of even a small amount of ethylbenzene may cause severe injury because its low viscosity and surface tension cause it to spread over a large surface of pulmonary tissue (HSDB 1990).

Ethylebenzene produces an irritant effect from chronic inhalation at 100 ppm/8 h (Patty 1981, as cited in HSDB).

It has been shown that the concentration of 1 mg/L and even 0.1 mg/L of ethylbenzene may be dangerous and may produce functional and organic disturbances (i.e., nervous system disorders, toxic hepatitis and upper respiratory tract complaints). Concentrations as low as 0.01 mg/L may lead to inflammation of upper respiratory tract mucosa (HSDB 1990).

Ethylbenzene vapor has a transient irritant effect on human eyes at 200 ppm in air. At 1000 ppm on the first exposure, it is very irritating and caused tearing, but tolerance rapidly develops. At 2000 ppm, eye irritation and lacrimation are immediate and severe. Concentrations of 5000 ppm cause intolerable irritation of the eyes and nose (HSDB 1990).

Human Carcinogenicity - The weight-of-evidence classification for ethylbenzene is "D; not classifiable as to human carcinogenicity." Ethylbenzene is nonclassifiable because of lack of animal bioassays and human studies (IRIS 1990). No human carcinogenicity data are available.

Animal Carcinogenicity - There are no data for animal carcinogenicity assessment. NTP has plans to initiate a bioassay. Metabolism and excretion studies at 3.5, 35 and 350 mg/kg are to be conducted as well.

Carcinogenicity Supporting Data - The metabolic pathways for humans and rodents are different (Engstrom et al. 1984, as cited in IRIS 1990). Major metabolites in humans, mandelic acid and phenylglyoxylic acid, are minor metabolites in rats and rabbits (Kiese and Lenk 1974, as cited in IRIS 1990). The major animal metabolites were not detected in the urine of exposed workers (Engstrom et al. 1984, as cited in IRIS 1990).

Ethylbenzene at 0.4 mg/plate was not mutagenic for Salmonella strains TA98, TA1535, TA1537 and TA1538 with or without Aroclor 1254 induced rat liver homogenates (S) (Nestmann et al. 1980, as cited in IRIS 1990). Ethhylbenzene was shown to increase the mean number of sister chromatic exchanges in human whole blood lymphocyte culture at the highest dose examined without any metabolic activation system (Norppa and Vainio 1983, as cited in IRIS 1990).

Dean et al. (1985), as cited in IRIS 1990, used a battery of short-term tests including bacterial mutation assays, mitotic gene conversion in Saccharomyces cerevisiae JD1 in the presence and absence of S9, and chromosomal damage in a cultured rat liver cell line. Ethylbenzene was not mutagenic in the range of concentrations tested (0.2, 2, 20, 50, and 200 μg/plate) for S. typhimurium TA98, TA100, TA1535, TA1537, and TA1538 or for Escherichia coli WP2 and WPSuvrA. Ethylbenzene also showed no response in the S. cerevisiae JD1 gene conversion assay. In contrast, ethylbenzene hydroperoxide showed positive responses with E. coli WP2 at 200 μg/plate in the presence of S9 and an equally significant response with the gene conversion system of yeast (IRIS 1990).

2.2.11 Formaldehyde

Formaldehyde is a clear, water-white, very slightly acid gas or liquid with a pungent odor. It has a density of 1.0, and a flash point of (15% methanol-free) 122°F. Its molecular formula is CH_2O and it has a molecular weight of 30.03. It has a Hazard Rating of 3 (Sax and Lewis 1989). Formaldehyde has a vapor pressure of 10 mm at -88°C (Patty 1963, as cited in IRIS 1990). The TLV, on a TWA basis, is 1 ppm. The TLV, on a STEL basis, is 2 ppm (ACGIH 1988).

2.2.11.1 Summary

The weight-of-evidence classification for formaldehyde is "B1; probable human carcinogen," based on limited evidence in humans and sufficient evidence in animals.

Effects in women include menstrual disorders and secondary sterility (IARC 1972-1985, as cited in IRIS 1990). Solutions splashed in eyes have caused injuries ranging from severe, permanent congeal opacification and loss of vision to minor discomfort (Grand 1974, as cited in IRIS 1990).

Medical conditions generally are aggravated by exposure to formaldehyde. In people sensitized to formaldehyde, late asthmatic reactions may be provoked by brief exposures at approximately 3 ppm (Hendrick 1982, as cited in IRIS 1990).

In a study to analyze the relation of chronic respiratory symptoms and ventilatory function, levels of Peak Expiratory Flow Rates (PEFR) in children decreased linearly as formaldehyde concentration increased. In adults, only morning PEFR levels were related to formaldehyde exposure.

The inhalation slope factor is 4.5E-2/mg/kg/day. The inhalation unit risk is $1.3E-5/\mu g/m^3$. The extrapolation method used was the linearized multistage procedure using additional risk.

The main signs and symptoms of exposure are irritation of the eyes, nose and throat; tearing; cough; bronchospasm; pulmonary irritation; and dermatitis (Proctor and Hughes 1978, as cited in IRIS 1990). Severe pain, vomiting and diarrhea result from ingestion. After absorption, formaldehyde depresses the central nervous system and symptoms similar to alcohol intoxication result. It can also cause a reduction in body temperature (Environment Canada 1982, as cited in IRIS 1990).

2.2.11.2 Health Effects

Noncarcinogenic Effects - A risk assessment for formaldehyde is under review by an EPA work group.

Effects in women include menstrual disorders and secondary sterility (IARC 1972-1985, as cited in IRIS 1990). Solutions splashed in eyes have caused injuries ranging from severe, permanent congeal opacification and loss of vision to minor discomfort (Grand 1974, as cited in IRIS 1990).

Medical conditions generally are aggravated by exposure to formaldehyde. In people sensitized to formaldehyde, late asthmatic reactions may be provoked by brief exposures at approximately 3 ppm (Hendrick 1982, as cited in IRIS 1990).

Eight symptomatic individuals chronically exposed to indoor formaldehyde at low concentrations (0.07-0.55 ppm) were compared to eight nonexposed subjects with respect to 1) presence of IgG and IgE antibodies to formaldehyde conjugated to human serum albumin (F-HSA); 2) the percentage of venous bolls T- and B-cells by E- and EAC-resetting; and 3) the ability of T- and B-cells to undergo mitogen (phytohemogglutin and pokeweed) stimulated blastogenesis as measured by the incorporation of tritiated thymidine. Anti-F-HSA, IgG, but not IgE, antibodies were detected in the sera of the eight exposed subjects; none were found in seven of the controls. T-lymphocytes were decreased in the exposed (48%) compared to the control (65.9%) subjects (P,0.01). B-cells were 12.6% (exposed group) and 14.75% (controls) (P,0.05). The incorporation of labeled thymidine by T-cells (phytohemagglutin) was decreased: 17,882 cpm (exposed group) and 28.576 cpm (controls) (p,0.01). T- and B-cell blastogenesis (poleweed) was 9,698 cpm (exposed group) and 11,219 (controls) (p,0.1) (HSDB 1990).

In a study to analyze the relation of chronic respiratory symptoms and ventilatory function, levels of PEFR in children decreased linearly as formaldehyde concentration increased. In adults, only morning PEFR levels were related to formaldehyde exposure. In nonsmokers, this effect was linear and much smaller than that found in children, even with concentrations of 100 parts per billion (ppb), the decrement due to this exposure was barely 1% of the normal morning value. In smokers the relation was quadratic and the decrease in PEFR with formaldehyde was found only for concentrations over 40 ppb. Data indicate an increased prevalence of chronic respiratory disease, chronic bronchitis or asthma, in children 6-15 years of ages living in houses with formaldehyde concentrations between 60 and 140 ppb. No threshold level was found for concentrations affecting ventilatory function (Krzyzanowski, Quickenboss, and Lebowitz 1989).

Human Carcinogenicity - The weight-of-evidence classification for formaldehyde is "Bl; probable human carcinogen," based on limited evidence in humans, and sufficient evidence in animals. Human data include nine studies that show statistically significant associations between site-specific respiratory neoplasms and exposure to formaldehyde or formaldehyde-containing products. An increased incidence of nasal squamous cell carcinomas was

observed in long-term inhalation studies in rats and in mice. The classification is supported by in vitro genotoxicity data and structural relationships
to other carcinogenic aldehydes such as acetaldehyde (IRIS 1990).

The human carcinogenicity data are limited. At least 28 epidemiologic
studies have been carried out relevant to formaldehyde. Among these, two
cohort (Blair et al. 1986, 1987; Stayner et al. 1988, as cited in IRIS 1990)
and one case-control (Vaughn et al. 1986a,b, as cited in IRIS 1990) were well
conducted and specifically designed to detect small to moderate increases in
formaldehyde-associated human risks. Blair et al. studied workers at ten
plants in some way exposed to formaldehyde (largely through resin formation)
and observed significant excesses in lung and nasopharyngeal cancer deaths.
Despite a lack of significant trends with increasing intensity or cumulative
formaldehyde exposure, lung cancer mortality was significantly elevated in
analyses with or without a 20-year latency allowance. No explicit control was
made for smoking status. Stayner et al. reported statistically significant
excesses in mortality from buccal cavity tumors among formaldehyde-exposed
garment workers. The highest mortality was for workers with long employment
duration (exposure) and follow-up period (latency). The Vaughn et al. nasal
and pharngeal cancer case-control study examined occupational and residential
exposures, controlling for smoking and alcohol consumption in analysis. It
showed a significant association between nasopharyngeal cancer and having
lived 10 or more years in a mobile home. People for whom this association was
drawn had lived in mobile homes built in the 1950s to 1970s, a period of
increasing formaldehyde-resin usage, but no measurements were directly available or used (IRIS 1990).

The 25 other reviewed studies had limited ability to detect small to
moderate increases in formaldehyde risks because of small sample sizes, small
numbers of observed site-specific deaths, and insufficient follow-up. Even
with these potential limitations, 6 of the 25 studies (Acheson et al. 1984;
Hardell et al. 1982; Hayes et al. 1986; Liebling et al. 1984; Olsen et al.
1984; Stayner et al. 1985, all as cited in IRIS 1990) reported significant
associations between excess site-specific respiratory (lung, buccal cavity and
pharyngeal) cancers and exposure to formaldehyde. Some of these studies

looked at such potential confounders as wood-dust exposure in greater detail; their resolution could not discern sinonasal cancer incidence excesses of the size predicted. Others (Liebling et al. 1984; Stayner et al. 1985, as cited in IRIS 1990) overlapped the three first-mentioned studies, the improved design and nonoverlapping portions of the later studies (Blair et al. 1986; Stayner et al. 1988, as cited in IRIS 1990) reinforce the conclusions of the earlier studies. Analysis of the remaining 19 studies indicates a possibility that observed leukemia and neoplasms of the brain and colon may be associated with formaldehyde exposure. The biological support for such postulates, however, has not yet been demonstrated. Although the common exposure in all of these studies was formaldehyde, the epidemiologic evidence is categorized as "limited" primarily because of possible exposures to other agents. Such exposures could have contributed to the findings of excess cancers (IRIS 1990).

Animal Carcinogenicity - The animal carcinogenicity data are considered sufficient. Consequences of inhalation exposure to formaldehyde have been studied in rats, mice, hamsters, and monkeys. The principal evidence comes from positive studies in both sexes of two strains of rats (Kerns et al. 1983; Albert et al. 1982; Tobe et al. 1985, all as cited in IRIS 1990) and males of one strain of mice (Kerns et al. 1983, as cited in IRIS 1990), all showing squamous cell carcinomas (IRIS 1990).

For Chemical Industry Institute of Toxicology (CIIT), Kerns et al. (1983), as cited in IRIS 1990, exposed about 120 animals per sex per species (Fischer 344 rats and B6C3F1 mice) to 0, 2, 5.6 or 14.3 ppm, 6 h/day, 5 day/wk for 24 months. Sacrifices of 5 per group were carried out at 6 and 12 months and 20 per group were killed at 18 months. At 24 and 27 months, the number sacrificed is unclear. The studies were terminated at 30 months. From the 12th month on, male and female rats in the highest dose groups showed mortality significantly increased over controls. In the 5.6 ppm group, male rats showed a significant increase from 17 months on. Female mice showed generally comparable survival across dose groups, as did male mice, but the male mice as a whole had survival decreased by housing-related problems. Squamous cell carcinomas were seen in the nasal cavities of 51 males and 52 females of the 117 male and 115 female rats at 14.3 ppm evaluated by experiments (as many as 35 had been identified in males by month 18 based on contemporaneous EPA analysis notes and Kerns Chart 8). Two rats (one male, one female) of 119 males and 116 females examined at 5.6 ppm showed squamous cell carcinomas of the nasal cavity. No such tumors were seen at 0 or 2 ppm. Polupoid adenomas of the nasal mucosa were seen in rats at all doses in a significant dose-related trend, albeit one that falls off after a peak. Among the mice, squamous cell carcinomas were seen in two males at 14.3 ppm. No other lesions were noteworthy.

Sellakumar et al. (1985), as cited in IRIS 1990, exposed male Sprague-Dawley rats, 100/group, 6 h/day, 5 day/wk for a lifetime, to a premix of 10 ppm HCL and 14 ppm formaldehyde, a combined exposure wherein HCl and formaldehyde were administered simultaneously but separately, 14 ppm formaldehyde alone, 10 ppm HCl alone, and an air control. An equal number of rats received no treatment. HCl was administered to determine whether tumor response was enhanced by an additional irritant effect or by the combining of formaldehyde and HCl to form bis-(chloromethyl)ether (BCME). Groups receiving formaldehyde alone or mixed with HCl were observed to have increased nasal squamous cell carcinomas; those without formaldehyde were free of carcinomas and other tumors (0/99 in each group), although rhinitis and hyperplasia were of comparable incidence (IRIS 1990).

A 28-month study of male Fischer 344 rats (begun about 2 weeks younger than those in Kerns et al. 1983; as cited in IRIS 1990) was carried out by Tobe et al. (1985), as cited in IRIS 1990. Groups of 32 rats were exposed 6 h/day, 5 day/wk to 15 ppm formaldehyde, 2, 0.3, or 3.3 ppm aqueous solution methanol (vehicle); an unexposed group of 32 was raised concurrently. Groups were compared to the methanol control and 12-, 18- and 24-month sacrifices were performed. Mortality was highest in the 15 ppm group, reaching 88% by month 24 when exposure to this group ceased. In the other groups, mortality by 28 months ranged from 60% (unexposed) to 32% (0.3 ppm). Squamous cell carcinomas were seen in 14/27 rats at 15 ppm surviving past 12 months and in no other rats (control incidence was 0/27). No polypoid adenomas were observed. Rhinitis and hyperplasia, while most prevalent at 15 ppm, were seen at increasing incidence across doses (IRIS 1990).

While these three rodent studies are principal in the weight of evidence, inhalation studies have been carried out in other strains and species. Dalbey (1982), as cited in IRIS (1990), as part of a promotion experiment, exposed male Syrian golden hamsters to 10 ppm formaldehyde five times per week, 5 hours per day throughout their lifetimes, with 132 animals as untreated controls. While survival time was significantly reduced in the treated group, no tumors were observed in either group. Rusch et al. (1983), as cited in IRIS (1990), carried out a one-half year toxicity study in six male cynomolgus monkeys, 40 F344 (10 male and 10 female) rats and 20 Syrian golden hamsters (20 male and 20 female) with higher than 2.95 ppm, with corresponding controls. The short duration of the assay, the small sample sizes and, possibly, the concentrations tested, limited the sensitivity of the assay to discern tumors. In the highest dose group in both rats and monkeys, incidence of squamous metaplasia/hyperplasia of the nasal turbinates was significantly elevated (IRIS 1990).

Carcinogenicity Supporting Data - Mutagenic activity of formaldehyde has been demonstrated in viruses, Escherichia coli, Pseudomonas florescence, Salmonella typhimurium and certain strains of yeast, fungi, Drosophila, grasshopper and mammalian cells (Ulsamer et al. 1984, as cited in IRIS 1990). Formaldehyde has been shown to cause gene mutations, single strand breaks in DNA, DNA-protein crosslinks, sister chromatic exchanges and chromosomal aberrations. Formaldehyde produces in vitro transformation in BALB/c 3T3 mouse cells, BHK21 hamster cells and C3H-10T1/2 mouse cells, enhances the transformation of Syrian hamster embryo cells by SA7 adenovirus, and inhibits DNA repair (Consensus Workshop of formaldehyde 1984, as cited in IRIS 1990).

Quantitative Carcinogenic Risk - The inhalation slope factor is 4.5E-2/ mg/kg/day. The inhalation unit risk is $1.3E-5/\mu g/m^3$. The extrapolation method used was the linearized multistage procedure using additional risk. Table 2.8 shows air concentrations at specified risk levels (IRIS 1990).

TABLE 2.8.   Formaldehyde Air Concentrations at
             Specified Risk Levels

Risk Level	Concentration
E-4 (1 in 10,000)	$8/\mu g/m^3$
E-5 (1 in 100,000)	$8E-1/\mu g/m^3$
E-6 (1 in 1,000,000)	$8E-2/\mu g/m^3$

Dose-Response Data - Inhalation dose-response data for carcinogenicity
are provided in IRIS (1990). The test animal was the male F344 rat;
scientists looked at squamous cell carcinoma. Table 2.9 shows the dose-
response data (IRIS 1990).

Rats that died prior to appearance of the first squamous cell carcinoma
at 11 months were not consider at risk. Those sacrificed at 12 and 18 months
were treated as though they would have responded in the same proportion as
rats remaining alive at the respective sacrifice times. Those living beyond
24 months were included with animals sacrificed at 24 months. From the estimates of the probability of death with tumor within 24 months and its variance, the number of animals at risk and the number with tumors were derived
for a 24-month study with no 12- or 18-month kills. These rounded numbers are
shown above and were used for significance tests and modeling. The unit risk
should not be used if the air concentration exceeds $8E+2$ $\mu g/m^3$, because above
this concentration, the slope factor may differ from that stated (IRIS 1990).

The experimental range is close to expected human exposures. Estimated
lifetime excess risks from six epidemiologic studies are close to upper bound
risks based on animal data (usually within 1 order of magnitude for four types
of estimated occupational and residential exposure). Animal-based estimates
derived using time in the model were similar but would have required the use
of more assumptions in calculation. Three non-zero doses were used in addition to controls in the study on which calculations are based, with a large
number of animals per group. Male and female incidences were close throughout
the exposure groups (IRIS 1990).

Toxicity Ranges - The probable oral lethal dose for humans is
0.5-5 g/kg, or between 1 ounce and 1 pint for a 150-lb person (Gosselin 1976,

TABLE 2.9. Formaldehyde Inhalation Exposure
             Dose-Response Data

Dose		
Administered (mg/kg/day)	Human Equivalent (mg/kg/day)	Tumor Incidence
Male		
0	0	0/156
2	2	0/159
5.6	5.6	2/153
14.3	14.3	94/140

as cited in IRIS 1990). Below 1 ppm, the odor of formaldehyde is perceptible to most people. At 2-3 ppm, mild tingling of the eyes occurs. At 4-5 ppm, increased discomfort with mild lacrimation occurs. At 10 ppm, profuse lacrimation occurs and can be withstood only for a few minutes. At 10-20 ppm, breathing difficulties, cough, and severe burning of nose and throat occur. At 50-100 ppm, acute irritation of respiratory tract occurs and very serious injury are likely. There is a delayed sensitization dermatitis (Proctor and Hughes 1978, as cited in IRIS 1990).

Table 2.10 shows the acute health effects from a Cleveland health effects study (Nied and TerKonda 1983).

Symptoms - The main signs and symptoms of exposure are irritation of eyes, nose and throat, tearing, cough, bronchospasm, pulmonary irritation, dermatitis (Proctor and Hughes 1978, as cited in IRIS 1990). Severe pain, vomiting and diarrhea result from ingestion. After absorption, formaldehyde depresses the central nervous system and symptoms similar to alcohol intoxication. It can also cause a reduction in body temperature (Environment Canada 1982, as cited in IRIS 1990).

2.2.12 Methylethylketone

Methylethylketone is a colorless liquid with an acetone-like odor. It has a density of 0.80615 and a flash point of 22°F. Its molecular formula is C_4H_8O and its molecular weight is 72.12. Methylethylketone has a vapor

TABLE 2.10. Formaldehyde Concentrations from Cleveland
Health Effects Study

Health Effects Reported	Average Formaldehyde Concentration (ppm)
No effects	0.02 - 0.26
Eye irritation	0.02 - 0.38
Nose and mucous membrane irritation	0.02 - 0.38
Neurophysiological effects	0.02 - 0.38
Upper airway irritation	0.03 - 0.38
Lower airway irritation	0.06 - 0.11
Dermatitis	0.04 - 0.14
Nausea and vomiting	0.03 - 0.13

pressure of 71.2 mm at 20°C (Sax and Lewis 1989). The TLV, on a TWA basis, for methylethylketone is 200 ppm. The TLV, on a STEL basis, is 300 ppm (ACGIH 1988). The most common synonym for methylethylketone is 2-butanone, with others such as MEK, ethylmethylketone, and methylacetone commonly used.

2.2.12.1 Summary

Methylethylketone is moderately toxic by ingestion, skin contact and intraperitoneal routes. Human systemic effects by inhalation include conjunctive irritation and unspecified effects on the nose and respiratory system. It is considered a strong irritant with human eye irritation at 350 ppm. Methylethylketone affects the peripheral nervous system and central nervous system. It has a Hazard Rating of 3 (Sax and Lewis 1989).

The RfDo is 5E-2mg/kg/day with a critical effect statement that no adverse effects were observed. The NOAEL listed for methylethylketone is 235 ppm, and the LOAEL listed is 130.5 mg/kg/day (IRIS 1990).

The weight-of-evidence classification for methylethylketone is "D; not classifiable as to human carcinogenicity." This classification is based on the fact that there are no human carcinogenicity data available and the animal data are considered inadequate (IRIS 1990).

2.2.12.2 Health Effects

General Toxicity Information - Methylethylketone is moderately toxic by ingestion, skin contact and intraperitoneal routes. Human systemic effects by inhalation include conjunctiva irritation and unspecified effects on the nose and respiratory system. Methylethylketone is an experimental teratogen. It has experimental reproductive effects. It is considered a strong irritant with human eye irritation at 350 ppm. Methylethylketone affects the peripheral nervous system and central nervous system (Sax and Lewis 1989).

Noncarcinogenic Effects - The RfD is based on the assumption that thresholds exist for certain toxic effects such as cellular necrosis, but may not exist for other toxic effects such as carcinogenicity. In general, RfD is an estimate of a daily exposure to the human population (including sensitive subgroups) that is likely to be without an appreciable risk of deleterious effects during a lifetime. An RfDo is available for methylethylketone. The RfDo is 5E-2mg/kg/day with a critical effect statement that no adverse effects were observed. The NOAEL listed for methylethylketone is 235 ppm, and the LOAEL listed is 130.5 mg/kg/day (IRIS 1990).

Adequate chronic toxicity testing has not been performed with methylethylketone. Although several more recent subchronic studies have been conducted (Freddi et al. 1982; Cavender et al. 1983; Takeuchi et al. 1983, all as cited in IRIS 1990), only the NOAEL of LaBelle and Brieger (1955), as cited in IRIS 1990, provides the lowest and most protective dose for deriving an RfD. In this study, 25 rats were exposed to 235 ppm of methylethylketone for 7 h/day, 5 day/wk for 12 weeks. No effects were observed, but only a few parameters were measured. Methylethylketone has also been tested for teratogencity (Schwetz et al. 1974; Deacon et al. 1981, as cited in IRIS 1990), and the observed LOAELs for fetotoxicity were higher than the NOAELs of LaBelle and Breiger (1955), as cited in IRIS 1990. The route extrapolation introduced uncertainty because of differences in pharmacokinetic parameters, notably absorption and elimination (IRIS 1990).

The study is given medium to low confidence because only 25 rats were exposed to only one dose, and the sex, strain, and amount of control animals were low because four different studies lent some support to the chosen NOAEL. Confidence in the RfD can also be considered medium to low (IRIS 1990).

Human Carcinogenicity - The weight-of-evidence classification for methylethylketone is "D; not classifiable as to human carcinogenicity." This classification is based on the fact that there are no human carcinogenicity data available and the animal data are considered inadequate.

Animal Carcinogenicity - The animal carcinogenicity data are considered inadequate. No data were available to assess the carcinogenic potential of methylethylketone by the oral or inhalation routes. In a skin carcinogenesis study, two groups of 10 male C3H/He mice received dermal applications of 50 mg of a solution containing 25 or 29% methylethylketone in 70% dodecylbenzene twice a week for 1 year. No skin tumors developed in the group of mice treated with 25% methylethylketone. After 27 weeks, a single skin tumor developed in 1 of 10 mice receiving 29% methylethylketone (Horton et al. 1965, as cited in IRIS 1990).

Carcinogenicity Supporting Data - methylethylketone was not mutagenic for Salmonella typhimurium strains TA98, TA100, TA1535, or TA1537 with or without rat hepatic homogenates (Florin et al. 1980; Douglas et al. 1980; as cited in IRIS 1990). Methylethylketone induced aneuploidy in the diploid D61, M strain of Saccharomyces cerevisiae (Zimmermann et al. 1985; as cited in IRIS 1990). Low levels of methylethylketone combined with low levels of nocodazile (another inducer of aneuploidy) also produced significantly elevated levels of aneuploidy in the system (Mayer and Goin 1987, as cited in IRIS 1990).

Quantitative Estimate of Carcinogenic Risk - No carcinogenic risk estimates are available for either the inhalation or oral routes of exposure.

2.2.13 Octane

Octane is a clear liquid with a density of 0.7036 and a flash point of 56°F. Its molecular formula is C_8H_{18}, its molecular weight 114.26. Octane has a vapor pressure of 10mm at 19.2°C. It has a Hazard Rating of 3 (Sax and Lewis 1989). The TLV, on a TWA basis, for octane is 300 ppm. The TLV, on a STEL basis, is 375 ppm (ACGIH 1988).

2.2.13.1 <u>Summary</u>

The amount of toxicological information available for octane is very limited. No specific studies involving human toxicology of octane were found.

Direct aspiration into the lungs of paraffins with carbon numbers C_6 to C_{16} may cause chemical pneumonitis, pulmonary edema, and hemorrhaging (HSDB 1990).

Octane is rated moderately toxic. It has a probable oral lethal dose for humans of 0.5 to 5 g/kg [i.e., between 1 oz. and 1 pint (or 1 pound) for a 70 kg (150 lb.) person] (HSDB 1990).

Rabbits with acute intoxication by inhaling carbon monoxide alone or in association with octane at concentrations of 5,000 ppm were compared by the Physiogram Method. Hydrocarbons aggravated the cardiovascular phenomena caused by carbon monoxide, and although they are central excitants, they further degraded electroencephalogram (EEG) activity (HSDB 1990).

2.2.13.2 <u>Health Effects</u>

The amount of toxicological information available for octane is very limited. No specific studies involving human toxicology of octane were found.

<u>General Toxicological Effects</u> - Octane is rated moderately toxic. It has a probable oral lethal dose for humans of 0.5 to 5 g/kg [i.e., between 1 oz. and 1 pint (or 1 pound) for a 70 kg (150 lb.) person] (HSDB 1990).

There may be epileptiform seizures months after an acute episode of octane. Pathological examination of tissues from fatal cases of octane exposure gives evidence for widespread microhemorragic phenomena. Irritation of the upper and lower respiratory tract and visceral damage have also been described (HSDB 1990).

Direct aspiration into the lungs of paraffins with carbon numbers C_6 to C_{16} may cause chemical pneumonitis, pulmonary edema, and hemorrhaging (HSDB 1990).

Mice exposed at concentrations of 6,600 to 13,700 ppm octane demonstrated central nervous system depression within 30 to 90 minutes and respiratory arrest at 16,000 (1 of 4) to 32,000 ppm (4 of 4) in 5 to 3 minutes, respectively (isooctane). The authors stated that the central nervous system depressant concentration was 10,000 ppm, which put the concentration at 8,000 ppm and the fatal concentration at 13,500 ppm. From these data on acute toxic response, it can reasonably be inferred that octane is from 1.2 to 2 times more toxic than haptene (HSDB 1990).

Orally, octane may be more toxic than its lower homologous. If the material is aspirated into the lungs, it may cause rapid death from cardiac arrest, respiratory paralysis, and asphyxia. The potency of Octane is approximately that of haptene, but does not appear to exhibit the central nervous system effect as to the two lower homologous (HSDB 1990).

The minimal concentration that causes loss of righting reflexes in mice was 35 mg/L, and total loss of reflexes occurred at 50 mg/L. A concentration of 95% causes loss of reflexes in mice in 125 minutes; however, a concentration of up to 1.9% appears to be tolerated for 143 minutes, with reversible effects (HSDB 1990).

Some effects of acute exposure to octane vapors on schedule-controlled responding in mice were detected. Cumulative concentration-effect curves were determined by comparing responding before and during exposure. The concentrations were incrementally increased at 40-minute intervals until responding completely ceased. Octane did not generally decrease responding until concentrations of approximately 1,000 ppm were obtained. Responding progressively decreased with increasing concentrations of up to 7,000 ppm (HSDB 1990).

Toxicant Interactions - Rabbits with acute intoxication by inhaling carbon monoxide alone or in association with octane at concentrations of 5,000 ppm were compared using the Physiogram Method. Hydrocarbons aggravated the cardiovascular phenomena caused by carbon monoxide, and although they are central excitants, they further degraded EEG activity (HSDB 1990).

2.2.14 α-Pinene

α-Pinene is a liquid with an odor of turpentine. It has a density of 0.8592 and a flash point of 91°F. Its molecular formula is $C_{10}H_{16}$ and its molecular weight is 136.26. α-Pinene has a vapor pressure of 10 mm at 37.3°C. It has a Hazard Rating of 3 (Sax and Lewis 1989).

2.2.14.1 Summary

The amount of toxicological information available for α-pinene is very limited. No specific studies involving human toxicology of α-pinene were found.

α-Pinene, in significant quantities, is considered a deadly poison by inhalation and moderately toxic by ingestion (Sax and Lewis 1989). It is also considered an eye, mucous membrane, and severe skin irritant. It has an oral rat LD50 of 3700 mg/kg. Its inhalation LCLo concentrations for rats, mice, and guinea pigs are 625 $\mu g/m^3$, 364 $\mu g/m^3$, and 572 $\mu g/m^3$ respectively (Sax and Lewis 1989).

2.2.14.2 Health Effects

The amount of toxicological information available for α-pinene is very limited. No specific studies involving human toxicology of α-pinene were found.

General Toxicity Information - α-Pinene, in significant quantities, is considered a deadly poison by inhalation and moderately toxic by ingestion (Sax and Lewis 1989). It is also considered an eye, mucous membrane, and severe skin irritant. It has an oral rat LD50 of 3700 mg/kg. Its inhalation LCLo concentrations for rats, mice, and guinea pigs are 625 $\mu g/m^3$, 364 $\mu g/m^3$, and 572 $\mu g/m^3$, respectively (Sax and Lewis 1989).

α-Pinene irritates the skin and mucous membranes. It can cause skin eruption, gastrointestinal (GI) tract irritation, delirium, ataxia, kidney damage and coma. Inhalation can cause palpitation, dizziness, nervous disturbances, chest pain, bronchitic, nephritis. α-Pinene can be absorbed through the skin, lungs, and intestine. It can also cause benign skin tumors from chronic contact (HSDB 1990).

α-Pinene has essentially the same toxicity as turpentine. As little as 15 ml (1/2 oz.) has proved fatal to a child, but a few children have survived 2 and even 3 ounces. The mean lethal dose in an adult probably lies between 4 and 6 ounces (HSDB 1990).

Toxicity Ranges - In an adult a dose of 140 ml may be fatal. A dose of 15 ml was fatal in a 2-year-old child; however, benzene was present in the mixture. Children have ingested 2 to 3 ounces and survived (HSDB 1990).

Symptoms and General Effects - The following symptomatology is associated with α-pinene: 1) burning pain in the mouth and throat, abdominal pain, nausea, vomiting and occasionally diarrhea; 2) mild respiratory tract symptoms (e.g., coughing, choking, dyspnea and even cyanosis) and pulmonary edema and pneumonitis from aspiration or systemic absorption; 3) transient excitement, ataxia, delirium and finally stupor, which is the commonest severe symptom; occasionally occurring convulsions, usually not until several hours after ingestion, when they may interrupt a deep coma; 4) occasionally occurring painful urination, albuminuria, and hematuria; urine may have an odor resembling violets; renal lesions are usually transient, 5) an odor of turpentine on breath and in vomitus; 6) fever and tachycardia; and 7) death from respiratory failure (HSDB 1990).

2.2.15   Propylbenzene

Propylbenzene (isocumene) is a clear liquid that is insoluble in water, but miscible in alcohol and ether. It has a density of 0.862 and a flash point of 86°F. Its molecular formula is C_9H_{12} and its molecular weight is 120.21. It has a vapor pressure of 10 mm at 43.4°C. Its most common synonym is isocumene, with 1-phenylpropane and n-propylbenzene also common synonyms. It has a Hazard Rating of 3 (Sax and Lewis 1989).

2.2.15.1   Summary

The amount of toxicological information available for propylbenzene (isocumene) is very limited. No specific studies involving human toxicology of propylbenzene were found.

Propylbenzene is considered mildly toxic by ingestion and inhalation (Sax and Lewis 1989). The rat oral LD50 is 6040 mg/kg. The mouse inhalation LCLo is 20 g/m^3 (RTECS 1987).

2.2.15.2 Health Effects

The amount of toxicological information available for propylbenzene (Isocumene) is very limited. No specific studies involving human toxicology of propylbenzene were found.

General Toxicity Information - Propylbenzene is considered mildly toxic by ingestion and inhalation (Sax and Lewis 1989). The rat oral LD50 is 6040 mg/kg. The mouse inhalation LCLo is 20 g/m^3 (RTECS 1987).

RTECS indicates that the toxic effects of propylbenzene have not yet been reviewed.

2.2.16 Radon/Radon Daughters

Radon is a colorless, odorless, inert gas that is very dense. It has a density, as a gas at 1 atmosphere 0 degrees, of 9.73 g/L, and, as a liquid at boiling point, of 4.4. It has a Hazard Rating of 3 (Sax and Lewis 1989). Radon is a decay product of uranium and it, in turn, decays to form radioactive progeny that may attach to dust particles or remain unattached. If these progeny are inhaled, they can be drawn into the lungs, where they emit alpha energy and may cause lung cancer.

2.2.16.1 Summary

The EPA has withdrawn its carcinogen assessment for radon and its daughters in IRIS pending further review. A new carcinogen summary is in preparation by the CRAVE Work Group (IRIS 1990). Thus, information presented for radon/radon-daughters is subject to change as a result of the new carcinogen summary by EPA.

The major health effects in the experimental studies that simulate exposures in the uranium mine environment are pulmonary emphysema, pulmonary fibrosis (pneumoconiosis), lung cancer, and lifespan shortening. In general, pulmonary fibrosis, emphysema, and lifespan shortening are not produced to any significant extent until radon-daughter exposures exceed about 5,000 working level months (WLM). However, lung cancer is produced in these studies using rats exposed to radon at levels down to 20 WLM, which are typical, average, human, lifetime, environmental exposure levels. Respiratory carcinoma, therefore, is the most prominent health effect associated with radon-daughter exposures (DOE 1987).

Risk-projection models have been developed to estimate lifetime lung-cancer risk for radon exposure in underground miners and for environmental exposures. The range of lifetime risk coefficients for most models lies between 1 and 5 lung cancers per 10,000 persons exposed per WLM. A plausible range for environmental exposures at the 10- to 20-WLM lifetime exposure level is from 0 (if a threshold exists) to about 5 per 10,000 per WLM (if all environmental lung cancer in nonsmokers is radon-daughter related). When calculating an incremental risk from increased environmental exposure, the age at exposure should be considered. A person aged 60, if exposed to an elevated radon-daughter level for the remainder of life (an average of 24 years), will have a lower risk than an individual aged 30 exposed similarly for 24 years because of the potential for more years of risk expression. The National Council on Radiation Protection and Measurements (NCRP) has developed tables that allow age at exposure to be taken into account (NCRP 1984a, as cited in DOE 1987). In general, except for persons over age 60, the lifetime risk appears to be between 1 and 2 per 10,000 per WLM. Although this is a small risk, the large number of persons exposed in the United States yields a significant number of lung cancers (DOE 1987).

The lifetime risk coefficient data developed by the NCRP for environmental exposures of radon are 9.1E-3/WLM/year and 3.6E-3/pCi radon/L for exposure from infancy to death. The equivalent values for exposures of populations having an age distribution equal to that in the United States in 1975 are 5.6E-3 and 2.1E-3, respectively. The values for populations are somewhat smaller because the number of years at risk following radon exposure is reduced compared to exposures starting at infancy (DOE 1987).

The International Commission on Radiological Protection (ICRP) draft report on environmental exposures (ICRP 1985, as cited in DOE 1987) recommends a lifetime individual-risk coefficient of E-2/WLM/year. This is in very good agreement with the 9.1E-3/WLM/year recommended by the NCRP (NCRP 1984a, as cited in DOE 1987).

The data are not yet fully consistent nor developed, but most indicate
that tobacco smoke in conjunction with radon exposure acts mainly as a cancer
promoter. The U.S. uranium-miner data indicate that smokers have a shorter
induction-latent period and a higher incidence of lung cancer than nonsmokers.

2.2.16.2 Health Effects

General Toxicological Information - In animals exposed to radon
daughters, lesions observed in organs other than the lung are considered
spontaneous or only indirectly related to exposure. These observations are
paralleled in the human exposures where effects in organs other than the lung
are either not prominent or cannot unequivocally be associated with radon-
daughter exposures alone (i.e., in the absence of associated pollutants
typically found in the mine environment, such as ore dust, silica, diesel
exhaust, and cigarette smoke). The major health effects in the experimental
studies that simulate exposures in the mine environment are pulmonary
emphysema, pulmonary fibrosis (pneumoconiosis), lung cancer, and lifespan
shortening. In general, pulmonary fibrosis, emphysema, and lifespan shortening are not produced to any significant extent until radon-daughter exposures
exceed about 5,000 WLM. However, lung cancer is produced in these studies
using rats exposed to radon down to levels of 20 WLM, which are typical
average human lifetime environmental exposure levels. Respiratory carcinoma,
therefore, is the most prominent health effect associated with radon-daughter
exposures (DOE 1987).

Human Carcinogenicity - The EPA has withdrawn its carcinogen assessment
for radon and its daughters in IRIS following further review. A new
carcinogen summary is in preparation by the CRAVE Work Group (IRIS 1990).

In most of the large epidemiological studies conducted to date, miners
with cumulative radon-daughter exposures somewhat below about 100 WLM indicate
excess lung cancer mortality. There are four major studies where a dose
response can be inferred (Lundin et al. 1971; National Academy of Sciences
1980; Muller et al. 1983, 1985; Sevc et al. 1976; Radford and Renard 1984, all
as cited in DOE 1987). These studies (followed since 1950) include 3,362 U.S.
underground uranium miners whose exposures range from 60 to 7000 WLM (average,
800); 15,984 Canadian uranium miners with upper-estimate exposures ranging

from 5 to 510 WLM (average, 74); 2,400 Czechoslovakian miners followed since 1948 with an exposure range from 72 to 716 WLM (average, 200); and 1,415 Swedish iron miners born between 1880 and 1919 who were alive in 1930 and exposed from 27 to 218 WLM (average, 80). To date, from 3% to 8% of the miners in the studies have developed lung cancer, mostly bronchogenic, attributable to radon-daughter exposures (i.e., above that expected from smoking or other causes alone) (DOE 1987).

The cell type of bronchogenic carcinoma does not clearly define the lung cancer etiology. Small-cell or oat-cell carcinoma is the earliest to appear, but all forms are increased by radon-daughter exposure. In underground miners with the longest latent intervals, epidermoid carcinoma appeared dominant. The tumor cell type has been shown to vary with many parameters, including smoking quantity, age at first underground exposure, and latent interval. The percentage of small-cell carcinoma decreased, and the percentage of epidermoid carcinoma increased with latent interval (Saccomanno et al. 1982, as cited in DOE 1987). In view of the numerous cell types involved, documenting tumor etiology by type seems unlikely (DOE 1987).

Quantitative Estimates of Carcinogen Risk - Respiratory-cancer risk projection has generally followed either an absolute or a relative risk model (NAS 1980; Thomas and McNeill 1982; Whittemore and McMillan 1983; NCRP 1984a; ICRP 1985, all as cited in DOE 1987). Lung cancer rarely appears before age 40, regardless of age at exposure, and never before a minimum latent interval following exposure from 5 to 10 years. Most absolute models are modified to include these basic features and an average annual rate of appearance. In a relative risk model, lifetime risk from radon-daughter exposure is directly proportional to the natural occurrence of lung cancer. Nonsmokers have approximately one-tenth the lung-cancer rate of smokers, and nonsmoking women have a lower rate than nonsmoking men. Thus, radon-daughter-related lung cancer could ultimately be dependent on sex, lifespan (which leads to overall higher lifetime lung-cancer risk), and smoking history (DOE 1987).

Not enough is known concerning the prediction accuracy of the various projection models when applied to specific mining groups. It is also possible that neither absolute nor relative risk models will generally describe the

miners' risk. The various mining cohorts under study will not go to closure
for about 20 more years; until then, the time course of lung-cancer development in humans for radon-daughter exposures will not be accurately known.

Underground miner data strictly allow the estimation of risk coefficients for radon-daughter exposures for adult males in the mine environment.
These data can be extrapolated to population exposures on the assumptions that
1) lung cancer is proportional to lung dose, and 2) the influence of associated pollutant exposures in the mine environment is negligible or is not
significantly different from the influence of associated pollutant exposures
in indoor air. The uncertainties in this extrapolation include uncertainties
in the following areas: 1) exposure levels of miners, 2) risk-projection,
3) radon-daughter dose to extrapolate adult-male data to populations (other
age groups, females, etc.), and 4) influence of co-pollutant exposures. However, despite the uncertainties and assumptions, no better database exists for
inferring population risks. The animal data support the concept of linearity
with dose and the existence of elevated risk down to human-lifetime,
background-exposure levels (DOE 1987).

Risk-projection models have been developed to estimate lifetime lung-
cancer risk for exposure in underground miners and for environmental exposures. The range of lifetime risk coefficients for most models lies between
1 and 5 lung cancers per 10,000 persons exposed per WLM. A plausible range
for environmental exposures at the 10- to 20-WLM lifetime exposure level is
from 0 (if a threshold exists) to about 5 per 10,000 per WLM (if all
environmental lung cancer in nonsmokers is radon-daughter related). When
calculating an incremental risk from increased environmental exposure, the age
at exposure should be considered. A person aged 60, if exposed to an elevated
radon-daughter level for the remainder of life (an average of 24 years), will
have a lower risk than an individual aged 30 exposed similarly for 24 years
because of the potential for more years of risk expression. The NCRP has
developed tables that allow age at exposure to be taken into account (NCRP
1984a, as cited in DOE 1987). In general, except for persons over age 60, the
lifetime risk appears to be between 1 and 2 per 10,000 per WLM. Although this

is a small risk, the large number of persons exposed in the United States yields a significant number of lung cancers (DOE 1987).

The lifetime risk coefficient data developed by the NCRP for environmental exposures are 9.1E-3/WLM/year and 3.6E-3/pCi radon/L for exposure from infancy to death. The equivalent values for exposures of populations having an age distribution equal to that in the United States in 1975 are 5.6E-3 and 2.1E-3, respectively. The values for populations are somewhat smaller because the number of years at risk following radon exposure are reduced compared to exposures starting at infancy (DOE 1987).

The ICRP draft report on environmental exposures (ICRP 1985, as cited in DOE 1987) recommends a lifetime individual-risk coefficient of E-2/WLM/year. This is in very good agreement with the 9.1E-3/WLM/year recommended by the NCRP (NCRP 1984a, as cited in DOE 1987).

Toxicant Interactions - The data are not yet fully consistent nor developed, but most indicate that tobacco smoke in conjunction with radon exposure acts mainly as a cancer promoter. The U.S. uranium-miner data indicate that smokers have a shorter induction-latent period and a higher incidence of lung cancer than nonsmokers. However, follow-up is relatively short in comparison to other mining groups, so the ultimate relationship is yet to be developed. Data from Swedish base-metal miners is quite different. These miners also show a shorter induction-latent period if they smoke; however, one study found a greater incidence of lung cancer among nonsmokers than smokers, and other studies found their incidence to be approximately equal (Axleson and Sundell 1978; Radford and Renard 1984, as cited in DOE 1987).

The main difference in the U.S. and Swedish studies is in the length of follow-up and, thus, the portion of lifespan over which data on deaths are collected. It is possible to resolve, at least partially, these various data by postulating that radiation exposures induced approximately the same finite numbers of cancers in both smokers and nonsmokers. Because these appear earlier in smokers, one sees a larger excess in nonsmokers if one looks particularly late in the development, and perhaps an approximately equal number of lung cancers when follow-up is over the lifespan of the two groups. Less

certainty exists about the relationship among smoking, radon-daughter exposures and lung cancer at low environmental radon-daughter exposure levels (DOE
1987).

The animal lung carcinoma data are also inconsistent but partially
explained by the temporal sequence of exposures. Smoke alternated the same
day with radon daughters can be "protective;" cumulative radon-daughter
exposures preceding smoke exposure produces synergism, while radon daughters
following the cumulative exposure to smoke produces no net effect over radon
daughters alone. The animal data, in general, support the concept that smoke
acts mainly as a promoter. Under some circumstances, it may promote more
mucus production and, therefore, shield sensitive cells from the radon-
daughter alpha radiation (DOE 1987).

2.2.17 Styrene

Styrene is a colorless, refractive, oily liquid. It has a density of
0.9074 and a flash point of 88°F. Its molecular formula is C_8H_8 and its
molecular weight is 104.16. It has a Hazard Rating of 3 (Sax and Lewis 1989).
The TLV, on a TWA basis, is 50 ppm. The TLV, on a STEL basis, is 100 ppm
(ACGIH 1988). Some of the common synonyms for styrene are vinylbenzene,
cinnamene, cinnamenal, cinnamol, ethenylbenzene, phenylethylene, phenethylene,
phyenylethylene, and styrene Monomer.

2.2.17.1 Summary

Styrene is irritating to eyes, the respiratory and gastrointestinal
tracts, and skin. Central nervous system depression may occur. Peripheral
neuropathies have been reported. "Styrene sickness" is common. Changes in
psychoneurological functioning have been described with chronic exposure.
Styrene is questionably mutagenic and teratogenic (HSDB 1990).

The RfDo for styrene is 2E-1 mg/kg/day, with the critical effect on red
blood cells and the liver. The NOAEL is 200 mg/kg/day and the LOAEL is 400
mg/kg/day (IRIS 1990).

Styrene is among those substances evaluated by the EPA for evidence of
human carcinogenic potential. This does not imply that this chemical is

necessarily a carcinogen. The evaluation for styrene is under review by an inter-office Agency work group (IRIS 1990).

An elevated incidence of hematopoietic and lymphatic cancer has been reported for workers in the styrene-butadiene rubber industry (HSDB 1990).

Styrene in concentrations of 100 ppm in air causes mild irritation of the eyes and throat in 20 minutes, but seems acceptable for working conditions. At 375 ppm, not all people feel significant eye irritations in 15 minutes, but all have nasal irritation. Concentrations of 400 and 500 ppm cause irritation of eyes and nose, but can be tolerated (HSDB 1990).

Styrene air concentrations of 10,000 ppm are dangerous to life within 20 to 30 minutes. Concentrations of 2,500 ppm are dangerous to life within 8 hours. Exposure to Styrene vapors above 800 ppm is immediately irritating. Eye damage can occur. High concentrations can have a toxic and anesthetic effect (1 hour at 1,000 ppm is severely toxic). One hour at 5,000 ppm produces unconsciousness; 10,000 ppm for 30 minutes has caused death. Repeated and prolonged contact of the liquid with the skin can cause defatting and dermatitis (HSDB 1990)

2.2.17.2 Health Effects

Noncarcinogenic Effects - The RfD is based on the assumption that thresholds exist for certain toxic effects such as cellular necrosis, but may not exist for other toxic effects such as carcinogenicity. In general, the RfD is an estimate of a daily exposure to the human population (including sensitive subgroups) that is likely to be without an appreciable risk of deleterious effects during a lifetime. The RfDo for styrene is 2E-1 mg/kg/day, with the critical effect on red blood cells and the liver. The NOAEL is 200 mg/kg/day and the LOAEL is 400 mg/kg/day (IRIS 1990).

Four beagle dogs per sex were gavaged with doses of 0, 200, 400, or 600 mg styrene/kg bw/day in peanut oil for 560 days. No adverse effects were observed for dogs administered styrene at 200 mg/kg/day. Effects in the higher dose groups were increased numbers of Heinz bodies in the red blood cells, decreased packed cell volume and sporadic decreases in hemoglobin and red blood cell counts, and increased iron deposits and elevated numbers of

Heinz bodies in the livers. Marked individual variations in blood cell parameters were noted for animals at the same dose level. Other parameters examined were body weight, organ weights, urinalyses, and clinical chemistry (IRIS 1990).

Long-term studies (120 weeks) in rats and mice (Ponomarkow and Tomatis 1978, as cited in IRIS 1990) showed liver, kidney, and stomach lesions for rats (dosed weekly with styrene at 500 mg/kg) and no significant effects for mice (dosed weekly with 300 mg/kg). Rats receiving an average daily oral dose of 95 mg of styrene/kw bw for 185 days showed no adverse effect, while those receiving 285 or 475 mg/kg/day showed reduced growth and increased liver and kidney weights (Wolf et al. 1956, as cited in IRIS 1990). Other subchronic rat feeding studies found LOAELs in the 350-500 mg/kg/day range and NOAELs in the range of 100-400 mg/kg/day (IRIS 1990).

The lifetime studies in rats and mice (Ponomarkov and Tomatis 1978, as cited in IRIS 1990) are not appropriate for risk assessment of chronic toxicity because of the dosing schedule employed. The Wolf et al. study (1956), as cited in IRIS (1990), is of insufficient duration to be considered chronic (IRIS 1990).

The confidence in the RfDo is medium. The principal study is well done and the effect levels seem reasonable, but the small number of animals per sex prevents a higher confidence at this time. The database offers strong support, but lacks a bona fide fullterm chronic study; thus, it is also considered to have a medium confidence. Thus, the medium confidence in the RfDo follows. The risk assessment for the RfDi is under review by an EPA work group (IRIS 1990).

Styrene is irritating to the eyes, the respiratory and GI tracts, and skin. Central nervous system depression may occur. Peripheral neuropathies have been reported. "Styrene sickness" is common. Changes in psychoneurological functioning have been described with chronic exposure. Styrene is questionably mutagenic and teratogenic (HSDB 1990).

Workers in a factory producing polystyrene resins, where concentrations reached 200 ppm, revealed itching dermatitis in one case and erythematous papular dermatitis of forearms in two other cases (HSDB 1990).

Ten men aged 20-41 years old, who were occupationally exposed to styrene, showed increases in the rate of chromosomal aberrations in cultured lymphocytes for peripheral blood (11-26% compared with 3% or less among 5 non-exposed controls). Decondensation of chromatic and increased numbers of micronuclei and nuclear bridges were also observed (HSDB 1990).

In a case where 449 workers exposed to styrene were examined, prenarcotic symptoms, such as light-headedness, eye irritation and irritation of mucous membranes were significantly more frequent in a "high" exposure group than in a "low" exposure group. A distal hypesthesia of the legs occurred in 8.5% of the cases. The conduction velocities of both radial and peroneal nerves were less than normal in 18.8% and 16.4% of the workers, respectively. There was consistent decrement in peroneal nerve conduction velocity as the exposure to styrene continued, but no such relationship was observed for radial nerve conduction velocities (HSDB 1990).

Effects on the liver (e.g., increased serum bile acid and enhanced activity of plasma enzymes) and reproductive system (e.g., decreased frequency of births and increased frequency of spontaneous abortions in female workers) have been reported (HSDB 1990).

The main pathological findings of exposure to styrene are edema of the brain and lungs, epithelial necrosis of the renal tubules and hepatic dystrophy (HSDB 1990).

Human Carcinogenicity - Styrene is among those substances evaluated by the EPA for evidence of human carcinogenic potential. This does not imply that this chemical is necessarily a carcinogen. The evaluation for styrene is under review by an inter-office Agency work group (IRIS 1990).

An elevated incidence of hematopoietic and lymphatic cancer has been reported for workers in the styrene-butadiene rubber industry. The researcher examined the 10-year mortality history of 6,678 male rubber workers. The age-related mortality rate due to lymphatic and hematopoietic cancer was reported to be 4.4 times higher for workers with 2 years of experience and 5.6 times higher for workers with 5 years of experience compared to the general study population. For lymphatic leukemia, the age-adjusted mortality rate in synthetics plant workers was 2.9 times higher for workers with 2 years of experience and 3.7 times higher for workers with 5 years of experience, compared to the general study population (HSDB 1990).

Toxicity Ranges - Styrene in concentrations of 100 ppm in air causes mild irritation of the eyes and throat in 20 minutes, but seems acceptable for working conditions. At 375 ppm, not all people feel significant eye irritations in 15 minutes, but all have nasal irritation. Concentrations of 400 and 500 ppm cause irritation of eyes and nose, but can be tolerated (HSDB 1990).

Styrene air concentrations of 10,000 ppm are dangerous to life within 20 to 30 minutes. Concentrations of 2,500 ppm are dangerous to life within 8 hours. Exposure to styrene vapors above 800 ppm is immediately irritating. Eye damage can occur. High concentrations can have a toxic and anesthetic effect (1 hour at 1,000 ppm is severely toxic). One hour at 5,000 ppm produces unconsciousness; 10,000 ppm for 30 minutes has caused death. Repeated and prolonged contact of the liquid with the skin can cause defatting and dermatitis (HSDB 1990).

Symptoms and General Effects - "Styrene sickness" is not uncommon in industry after exposure to vapors or mists. Characteristic signs and symptoms include headache, fatigue, weakness, depression, and unsteadiness or feeling of drunkenness, and an abnormal EEG (HSDB 1990).

An acute ingestion or inhalation of styrene can have the following symptoms: 1) ingestion causes a burning sensation in the mouth and stomach, and causes nausea, vomiting and salivation; hematemesis may occur; 2) substernal pain, cough and hoarseness are described; 3) aspiration into the tracheobronchial tree, either during ingestion or subsequent to vomiting or eructation, is likely to produce a severe hemorrhagic pneumonitis; 4) in vapor exposures, a transient euphoria is sometimes observed; 5) headache, giddiness, vertigo, ataxia and tinnitus occur; 6) confusion, stupefaction and coma occur; 7) often associated with this coma are tremors, motor restlessness, hypertonus and hyperactive reflexes; 8) death occurs from respiratory failure or from sudden ventricular fibrillation; 9) skin contact with liquid may cause erythema and even blisters if the contact is prolonged. Hemorrhagic inflammatory lesions develop on mucous membranes in contact with liquid (HSDB 1990).

A chronic or repeated inhalation of styrene can have the following symptoms. Severe muscle weakness leads to limb paralysis, associated with hypokalemia from renal tubular acidosis. Cardiac arrhythmias often accompany the hypokalemia. Sensory function and tendon reflexes are not impaired. Gastrointestinal complaints, including abdominal pain, nausea, vomiting and hematemesis, occur but there are no significant abdominal tenderness or palpable abdominal masses (HSDB 1990).

2.2.18 Tetrachloroethylene

Tetrachloroethylene is a colorless liquid with a chloroform-like odor. It has a density of 1.6311 and no registered flash point. Its molecular formula is C_2Cl_4 and its molecular weight is 165.82. Tetrachloroethylene has a vapor pressure of 15.8 mm at 22°C. It has a Hazard Rating of 3 (Sax and Lewis 1989). The TLV, on a TWA basis, is 50 ppm. The TLV, on a STEL basis, is 200 ppm (ACGIH 1988). The most common synonym is perchloroethylene, with carbonbichloride, carbordichloride, PCE, PER, PERC, perchlor, PERK, 1,1,2,2-tetrachloroethylene, and tetrachloroethane the common synonyms.

2.2.18.1 Summary

Tetrachloroethylene is toxic by ingestion, inhalation, or dermal exposure. Vasodilation and malaise ("degreasers' flush") occur in workers who drink ethanol after exposure to trichloroethylene. Inhalation abuse of typewriter correction fluid has been reported (HSDB 1990).

The RfDo for terachloroethylene is 1E-2 mg/kg/day, with the critical effect on hepatotoxicity in mice, weight gain. The NOAEL is 20 mg/kg/day and the LOAEL is 100 mg/kg/day (IRIS 1990).

Excessive exposure to tetrachloroethylene has resulted in effects on the central nervous system, mucous membranes, eyes and skin, and to a lesser extent, to the lungs, liver, and kidneys. The effects most frequently noted have been on the nervous system. Unconsciousness, dizziness, headache, vertigo or light central nervous system depression have occurred in many instances after occupational exposures (HSDB 1990).

Several studies of the effects of prolonged exposure to perchloroethyl- ene vapors on human volunteers are available. Prolonged exposure to 200 ppm results in early signs of central nervous system depression, while there was no response in men or women reportedly exposed to 100 ppm for 7 h/day. Clinical studies indicate no liver or kidney effects at these levels, but massive exposure to concentrations causing unconsciousness have resulted in proteinuria and hematuria (HSDB 1990).

2.2.18.2 Health Effects

Noncarcinogenic Effects - The RfD is based on the assumption that thresholds exist for certain toxic effects such as cellular necrosis, but may not exist for other toxic effects such as carcinogenicity. In general, the RfD is an estimate of daily exposure to the human population (including sensitive subgroups) that is likely to be without an appreciable risk of deleterious effects during a lifetime. The RfDo for terachloroethylene is 1E-2 mg/kg/day, with the critical effect on hepatotoxicity in mice, weight gain. The NOAEL is 20 mg/kg/day and the LOAEL is 100 mg/kg/day (IRIS 1990).

Buben and O'Flaherty (1985), as cited in IRIS (1990), exposed Swiss-Cox mice to tetrachloroethylene in corn oil by gavage at doses of 0, 20, 100, 200, 500, 1500, and 2000 mg/kg at 5 day/wk for 6 weeks. Liver toxicity was evaluated by several parameters including liver weight/body weight, hepatic triglyceride concentration, DNA content, histopathological evaluation, and serum enzyme levels. Increased liver triglycerides were first observed in mice treated with 100 mg/kg. Liver weight/body weight ratios were signifi- cantly higher than controls for animals treated with 100 mg/kg. At higher doses, hepatotoxic effects included decreased DNA content, increased SGPT, decreased levels of glucose-6-phosphate (G6P) and hepatocellular necrosis, degeneration and polyploidy (IRIS 1990).

A NOEL of 14 mg/kg/day was established in a second study as well (Hayes et al. 1986; as cited in IRIS 1990). Groups of 20 Sprague-Dawley rats of both sexes were administered doses of 14, 400, or 1400 mg/kg/day in drinking water.

Males in the high-dose group and females in the two highest groups exhibited depressed body weights. Equivocal evidence of hepatotoxicity (increased liver and kidney weight/body weight ratios) were also observed at higher doses (IRIS 1990).

Other data support the findings of the principal studies. Exposure of mice and rats to tetrachloroethylene by gavage for 11 days caused hepato-toxicity (centrilobular swelling) at doses as low as 100 mg/kg/day in mice (Schaum et al. 1980, as cited in IRIS 1990). Mice were more sensitive to the effects of tetrachloroethylene exposure than were rats. Increased liver weight was observed in mice at 250 mg/kg, but rats did not exhibit these effects until doses of 1000 mg/kg/day were reached. Relative sensitivity to humans cannot be readily established, but the RfD of 1E-2 mg/kg/day is protective of the most mild effects observed in humans [diminished odor perception/modified Romberg test scores in volunteers exposed to 100 ppm for 7 hours; roughly equivalent to 20 mg/kg/day (Stewart et al. 1961, as cited in IRIS 1990)].

The principal studies are of short duration. Inhalation studies have been performed that indicate that the uncertainty factor of 10 is sufficient for extrapolation of the subchronic effect to its chronic equivalent. Liver enlargement and vacuolation of hepatocyte were found to be reversible lesions for mice exposed to low concentrations of tetrachloroethylene (Kjellstrand et al. 1984, as cited in IRIS 1990). In addition, elevated liver weight/body weight ratios observed in animals exposed to tetrachloroethylene for 30 days were similar to those in animals exposed for 120 days. Several chronic inhalation studies have also been performed (Carpenter 1937; NTP 1985; Rowe et al. 1952, as cited in IRIS 1990). None are inconsistent with a NOAEL of 14 mg/kg/day for tetrachloroethylene observed by Buben and O'Flaherty (1985) and Hayes et al. (1986), both as cited in IRIS (1990).

The confidence in the RfDo is medium. No one study combines the fea-tures desired for deriving an RfD: oral exposure, larger number of animals, multiple dose groups, testing in both sexes and chronic exposure. Confidence in the principal studies is low mainly because of the lack of complete histo-pathological examination at the NOAEL in the mouse study. The database is relatively complete but lacks studies of reproductive and teratology endpoints subsequent to oral exposure; thus, it receives a medium confidence rating, with the medium confidence rating for the RfDo following (IRIS 1990).

An RfDi is not available at this time.

Excessive exposure to tetrachloroethylene has resulted in effects on the central nervous system, mucous membranes, eyes and skin, and to a lesser extent, to the lungs, liver, and kidneys. The effects most frequently noted have been on the nervous system. Unconsciousness, dizziness, headache, vertigo or light central nervous system depression have occurred in many instances after occupational exposures (HSDB 1990).

Several studies of the effects of prolonged exposure to tetrachloro-ethylene vapors on human volunteers are available. Prolonged exposure to 200 ppm results in early signs of central nervous system depression, but there was no response in men or women reportedly exposed to 100 ppm for 7 h/day. Clinical studies indicate no liver or kidney effects at these levels, but massive exposure to concentrations causing unconsciousness have resulted in proteinuria and hematuria (HSDB 1990).

A 6-wk-old breast-fed infant had obstructive jaundice and hepatomegaly. Tetrachloroethylene was detected in the milk and blood. After discontinuance of breast-feeding, rapid clinical and biochemical improvement were noted (HSDB 1990).

After ingestion of 12-16 g of tetrachloroethylene, a 6-yr-old boy was admitted to the clinic in coma. In view of the high initial tetrachloro-ethylene blood level, hyperventilation therapy was performed. Under this therapeutic regimen, the clinical condition of the patient improved consider-ably. The tetrachloroethylene blood level profile that was determined under hyperventilation therapy could be computer-fitted to a two-compartment model. Elimination of tetrachloroethylene from the blood compartment occurred via a rapid and a slow process with half-lives of 30 minutes and 35 hours, respec-tively. These values compared favorably with the half-lives of 160 minutes and 33 hours under normal respiratory conditions. During hyperventilation therapy, the relative contribution to the fast elimination process increased from 70% for physiological minute volume to 99.9%. A minor fraction of the ingested dose was excreted with the urine (integral of 1% during the first 3 days). In contrast to previous results, trace amounts of unchanged tetrachloroethylene were detected in the urine in addition to Trichloracetic acid and trichloroethanol (HSDB 1990).

Alterations in liver function in persons exposed to unknown concentrations of tetrachloroethylene over extended (chronic) periods have been reported by a number of investigators (Coler and Rossmiller 1953; Franke and Eggeling 1969; Hughes 1954; Trense and Zimmerman 1969; Meckler and Phelps 1966; Larsen et al. 1977; Moeschlin 1965; Dumortier et al. 1964, all as cited in EPA 1985). Liver function parameters that have been altered as a result of excessive tetrachloroethylene exposure include sulfobromophthalein retention time, thymol turbidity, serum bilirubin, serum protein patterns, cephalin-cholesterol flocculation, serum alkaline phosphatase, SGOT, and serum lactic acid dehydrogenase (LDH). However, these parameters may result from other causes that are completely dissociated with tetrachloroethylene (EPA 1985).

Absorption - Studies have shown that absorption of tetrachloroethylene through the skin, from vapor exposure or from partial body immersion, is minimal in comparison to oral and inhalation routes of exposure. Tetrachloroethylene is rapidly and completely absorbed into the body from the GI tract, presumably because of its high lipid solubility. tetrachloroethylene in vapor form in air is readily absorbed through the lungs into blood by first-order diffusion processes. Pulmonary uptake of a volatile compound like tetrachloroethylene during inhalation exposure is largely determined by the ventilation rate (about 4 to 8 L/min for humans at rest), duration of exposure at a given air concentration, solubility in blood and other body tissues, and its metabolism. When body tissue concentrations (body burden) are at steady-state with inspired air concentration, the rate of uptake is equal to the rate of metabolism plus nonpulmonary excretion of tetrachloroethylene. Because there are no known significant routes of excretion of tetrachloroethylene except pulmonary and metabolism, the steady-state uptake rate approximates the metabolism rate (EPA 1985).

Human Carcinogenicity - Tetrachloroethylene is among the chemicals evaluated by the EPA for evidence of human carcinogenic potential. This does not imply that it is necessarily a carcinogen. The evaluation for tetra-chloroethylene is under review by an inter-office Agency work group. A risk assessment summary will be included in IRIS when the review is completed (IRIS 1990).

Toxicity Ranges - The estimated dose in humans to cause death is reported to be 3 to 5 ml/kg. An oral toxic dose for Tetrachloroethylene has not been established. The TLV for both is 50 ppm (HSDB 1990).

Toxicant Interactions - Stewart et al. (1977), as cited in EPA (1985), conjectured that repeated tetrachloroethylene exposure effects might be exacerbated by simultaneous administration of either alcohol or Diazepam. While both alcohol and Diazepam produced decrements in the various objective tests, the effects were no worse when these substances were combined with tetrachloroethylene exposure. It would appear that repeated 100 ppm tetrachloroethylene exposure is not close to the threshold for objective test effects (EPA 1985).

2.2.19 Toluene

Toluene is a colorless liquid with a benzol-like odor. It has a density of 0,866 and a flash point of 40°F. Its molecular formula is C_7H_8 and its molecular weight is 92.15. Toluene has a vapor pressure of 36.7 mm at 30°C (Sax and Lewis 1989). The TLV, on a TWA basis, is 100 ppm. The TLV, on a STEL basis, is 150 ppm (ACGIH 1988). Common synonyms for toluene are methylben-zene, methacide, methylbonzol, phenylmethane, and toluol.

2.2.19.1 Summary

The RfDo for toluene is 3E-1 mg/kg/day, with the critical effect on clinical chemistry and hematological parameters. It should be noted that the RfDo for toluene may change in the near future pending the outcome of a further review now being conducted by the Oral RfD Workgroup of EPA. The NOAEL for toluene is 300 ppm. There is no LOAEL listed for toluene. (IRIS 1990).

The weight-of-evidence classification is "D; not classified." The basis for this classification is no human data and inadequate animal data. Toluene did not produce positive results in the majority of genotoxic assays (IRIS 1990).

2.2.19.2 Health Effects

Noncarcinogenic Effects - The RfD is based on the assumption that thresholds exist for certain toxic effects such as cellular necrosis, but may not exist for other toxic effects such as carcinogenicity. In general, the RfD is an estimate of a daily exposure to the human population (including sensitive subgroups) that is likely to be without an appreciable risk of deleterious effects during a lifetime. The RfDo for toluene is 3E-1 mg/kg/day, with the critical effect on clinical chemistry and hematological parameters. It should be noted that the RfDo for toluene may change in the near future pending the outcome of a further review being conducted by the Oral RfD Workgroup of EPA. The NOAEL for toluene is 300 ppm. There is no LOAEL listed for toluene. (IRIS 1990).

Toluene is most likely a potential source of respiratory hazard. The only chronic toxicity study on toluene was conducted for 24 months in male and female F344 rats (CIIT 1980, as cited in IRIS 1990). Toluene was administered by inhalation at 30, 100, or 300 ppm to 120 male and 120 female F344 rats for 6 h/day, 5 day/wk. The same number of animals (120 males and 120 females) was used as a control. Clinical chemistry, hematology and urinalysis testing was conducted at 18 and 24 months. All parameters measured at the termination of the study were normal except for a dose-related reduction in hematocrit values in females exposed to 100 and 300 ppm toluene. Based on these findings, a NOAEL of 300 ppm was derived. An oral RfD of 20 mg/day can be derived using route-to-route extrapolation. This was done by expanding the exposure from 6 h/day, 5 day/wk to continuous exposure and multiplying by 20 m^3/day and 0.5 to reflect a 50% absorption factor (IRIS 1990).

Subchronic inhalation and subchronic oral studies in both mice and rats support the chosen NOAEL (NTP 1981, 1982, as cited in IRIS 1990). Further-more, an oral study (Wolf et al. 1956, as cited in IRIS 1990) contains subchronic data in which no adverse effects of toluene were reported at the highest dose tested (590 mg/kd/day).

The confidence in the RfDo is medium. Confidence in the principal study is high because a large number of animals per sex were tested in each of three dose groups and many parameters were studied. Interim kills were performed. The database is rated medium because several studies support the chosen effect level. The confidence of the RfDo is no higher than medium because the critical study was by the inhalation route (IRIS 1990).

Human Carcinogenicity - The weight-of-evidence classification is "D; not classified." The basis for this classification is no human data and inadequate animal data. Toluene did not produce positive results in the majority of genotoxic assays (IRIS 1990).

Animal Carcinogenicity - A chronic (106-week) bioassay of toluene in F344 rats of both sexes reported no carcinogenic responses (CIIT 1980, as cited in IRIS 1990). A total of 960 rats were exposed by inhalation for 6 h/day, 5 day/wk to toluene at 0, 30, 100, or 300 ppm. Groups of 20981/sex/ dose were sacrificed at 18 months. Gross and microscopic examination of tissues and organs identified no increase in neoplastic tissue or tumor masses among treated rats when compared with controls. The study is considered inadequate because the highest dose administered was well below the MTD for toluene and because of the high incidence of lesions and pathological changes in the control animals (IRIS 1990).

Several studies have examined the carcinogenicity of toluene following repeated dermal applications. Toluene (dose not reported) applied to shaved interscapular skin of 54 male mice (strains A/He, C3HeB, SWR) throughout their lifetime (3 times weekly) produced no carcinogenic response (Poel 1963, as cited in IRIS 1990). One drop of toluene (about 6 ml) applied to the dorsal skin of 20 random-bred albino mice twice weekly for 50 weeks caused no skin papillomas or carcinomas after a 1-year latency period was allowed (Coombs et al. 1973, as cited in IRIS 1990). No increase in the incidence of skin or systemic tumors was demonstrated in male or female mice of three strains (CF, C3H, or CBaH) when toluene was applied to the back of 25 mice of each sex of each strain at 0.05-0.1 ml/mouse, twice weekly for 56 weeks (Doak et al. 1976, as cited in IRIS 1990). One skin papilloma and a single skin carcinoma were reported among a group of 30 mice treated dermally with one drop of 0.2% (w/v) solution toluene twice weekly, administered from droppers delivering 16-20 μL per drop for 72 weeks (Lijinsky and Garcia 1972, as cited in IRIS 1990). It is not reported whether evaporation of toluene from the skin was prevented during these studies (IRIS 1990).

Carcinogenicity Supporting Data - Toluene was found to be nonmutagenic in reverse mutation assays with S. typhimurium (Mortelmans and Riccio 1980; Nestmann et al. 1980; Bos et al. 1981; Litton Bionetics, Inc. 1981; Snow et al. 1981, as cited in IRIS 1990) and E. coli (Mortelmans and Riccio 1980, as cited in IRIS 1990), with and without metabolic activation. Toluene did not induce mitotic gene conversion (Litton Bionetics, Inc. 1981; Mortelmans and Riccio 1980, as cited in IRIS 1990) or mitotic crossing over (Mortelmans and Riccio 1980, as cited in IRIS 1990) in S. cerevisiae. Although Litton Bionetics, Inc. (1981), as cited in IRIS (1990), reported that toluene did not cause increased chromosomal aberrations in bone marrow cells, several Russian studies (Dobrokhtov 1972; Lyapkalo 1973, as cited in IRIS 1990) report toluene was effective in causing chromosal damage in bone marrow cells of rats. There was no evidence of chromosomal aberrations of blood lymphocytes of workers exposed to toluene only (Maki-Paakkanen et al. 1980; Forni et al. 1971, as cited in IRIS 1990), although a slight increase was noted in workers exposed to toluene and benzene (Forni et al. 1971; Funes-Craviota et al. 1977, as cited in IRIS 1990). This is supported by studies of cultured human lymphocytes exposed to toluene in vitro. No elevation of chromosomal aberrations or sister chromatic exchanges was observed (Gerner-Smidt and Friedrich 1978, as cited in IRIS 1990).

Quantitative Estimate of Carcinogen Risk - No estimate of the carcinogenic risk of toluene is available.

2.2.20 1,1,1-Trichloroethane

1,1,1-trichloroethane is a colorless liquid with a density of 1.3376 and no listed flash point. Its molecular formula is $C_2H_3Cl_3$ and its molecular weight is 133.40. 1,1,1-trichloroethane has a vapor pressure of 100 mm at 20.0°C. It has a Hazard Rating of 3 (Sax and Lewis 1989). The TLV, on a TWA basis, is 350 ppm. The TLV, on a STEL basis, is 450 ppm (ACGIH 1988). The most common synonym for 1,1,1-trichloroethane is methylchloroform, with aerothene TT, chlorothene, inhibisol, methyltrichloromethane, strobane, alpha-T, 1,1,1-TCE, and alpha-trichloroethane other common synonyms.

2.2.20.1 Summary

1,1,1-trichloroethane is one of the least toxic of the chlorinated hydrocarbons used as a solvent. Trichloroethane is a central nervous system and respiratory depressant and a skin and mucous membrane irritant (HSDB 1990).

The RfDo for 1,1,1-trichloroethane is 9E-2 mg/kg/day, with no critical adverse effects listed. The NOAEL is 500 ppm (air), and the LOAEL is 650 ppm (air) (IRIS 1990).

Men exposed to 900 to 1000 ppm of 1,1,1-trichloroethane experienced transient mild irritation and minimal impairment of coordination. Above 1700 ppm, this magnitude may induce headache and lassitude. Prolonged or repeated contact with skin results in transient erythema and slight irritation from the defatting action of the solvent. Some cases have also been observed in which repeated skin contact will cause a serious dermatitis characterized by skin cracking and infection from the defatting action of the solvent (HSDB 1990).

Four groups of workers (total number = 196) exposed for at least 5 years to average concentrations of 4, 25, 28 and 53 ppm methylchloroform were evaluated for adverse effects. Routine laboratory examinations, including peripheral hemograms, blood specific gravity and urinalysis, plus the sense-of-vibration test, were used. No consistent dose-related adverse effects were observed (HSDB 1990).

The estimated dose-response relationship for acute effects of single, short-term exposures, for concentrations above 5000 ppm, is the onset of central nervous system depression. Concentrations above 7,500 ppm, are possibly life threatening. A concentration of 2650-1900 ppm produces light-headedness and irritation of the throat. A concentration of 1000 ppm produces disturbance of equilibrium. Concentrations of 500 to 350 ppm produce slight changes in perception and obvious odor. A concentration of 100 ppm is the apparent odor threshold. It should be noted that these approximations are crude estimates and contain several areas of uncertainty (HSDB 1990).

2.2.20.2 Health Effects

Noncarcinogenic Effects - The RfD is based on the assumption that thresholds exist for certain toxic effects such as cellular necrosis, but may not exist for other toxic effects such as carcinogenicity. In general, the RfD is an estimate of a daily exposure to the human population (including sensitive subgroups) that is likely to be without an appreciable risk of deleterious effects during a lifetime. The RfDo for 1,1,1-trichloroethane is 9E-2 mg/kg/day, with no critical adverse effects listed. The NOAEL is 500 ppm (air) and the LOAEL is 650 ppm (air) (IRIS 1990).

Torkelson et al. (1958), as cited in (IRIS) 1990, exposed groups of rats, rabbits, guinea pigs and monkeys to 1,1,1-trichloroethane vapor at concentrations of 500, 1000, 2000, or 10,000 ppm. From these studies, it was determined that the female guinea pig was the most sensitive of the animals tested. At 500 ppm, groups of eight male and eight female guinea pigs showed no evidence of adverse effects compared with unexposed and air-exposed controls after exposure for 7 h/day, 5 day/wk for 6 months. Groups of five female guinea pigs exposed to 1,000 ppm 1,1,1-trichloroethane vapor 3 h/day, 5 day/wk for 3 months had fatty changes in the liver and statistically significant increased liver weights. Thus, this study defined a NOAEL of 500 ppm in guinea pigs (IRIS 1990).

Adams et al. (1950), as cited in IRIS (1990), subjected groups of 6-10 male and female guinea pigs to 650 ppm 1,1,1-trichloroethane vapor 7 h/day, 5 day/wk for 2 to 3 months. These animals exhibited a slight depression in weight gain compared with both air-exposed and unexposed controls, thereby establishing a LOAEL of 650 ppm in guinea pigs (IRIS 1990).

The 1,1,1-trichloroethane samples used by Torkelson et al. (1958), as cited in IRIS (1990), were found to be 94-97% pure while the samples used by Adams et al. (1950), as cited in IRIS (1990), had a purity of greater than or equal to 99% (IRIS 1990).

In addition, the effects of 1,1,1-trichloroethane vapor have been investigated in mice (Quast et al. 1984; McNutt et al. 1975, as cited in IRIS
1990), rats (Quast et al. 1978, as cited in IRIS 1990), and rabbits and dogs
(Pendergast et al. 1967, as cited in IRIS 1990). The only chronic oral exposure study was conducted by NCI (1977), as cited in IRIS (1990), in rats. The
observations from these studies and from Torkelson et al. (1958), as cited in
IRIS (1990), and Adams et al. (1950), as cited in IRIS (1990), are somewhat
inconsistent, thus making it difficult to draw conclusions about the dose
levels of 1,1,1-trichloroethane that result in adverse effects. For example,
inhalation exposure to 650 ppm by Adams et al. (1950), as cited in IRIS
(1990), was associated with slight growth retardation in guinea pigs. Further
review of this study indicates that 1500 ppm exposure also caused slight
growth retardation without causing any organ-specific adverse effects following 1 to 3 months exposure. These observations are in contrast with those of
Torkelson et al. (1958), as cited in IRIS (1990), who observed adverse effects
in the liver and lungs of guinea pigs exposed to 1,000 ppm for 90 days (IRIS
1990).

When published, the results and technical evaluation of recent inhalation studies in mice (Quast et al. 1984, as cited in IRIS 1990) and rats
conducted by Dow Chemical may be of greater value for the overall RfD
consideration for 1,1,1-trichloroethane (IRIS 1990).

The confidence in the RfDo is medium. Although studies by Adams et al.
(1950), as cited in IRIS (1990), and Torkelson et al. (1958), as cited in IRIS
(1990), used both sexes of several species, the number of animals at each dose
level was limited, the length of exposure varied with different dose levels
and few toxic endpoints were examined. Confidence in these studies is thus
considered low. The database is fairly comprehensive; however, results from
these studies are somewhat inconsistent and some of the more recent studies
have yet to be critically evaluated. Confidence in the database is, therefore, rated medium. Confidence in the RfDo can be considered medium to low
(IRIS 1990).

Men exposed to 900 to 1000 ppm of 1,1,1-trichloroethane experienced
transient mild irritation and minimal impairment of coordination. Above

1700 ppm, this magnitude may induce headaches and lassitude. Prolonged or repeated contact with skin results in transient erythema and slight irritation from the defatting action of the solvent. Some cases have also been observed in which repeated skin contact causes serious dermatitis characterized by skin cracking and infection from the defatting action of the solvent (HSDB 1990).

The effect of the solvent methylchloroform on psychophysiological functions, such as reaction time, perceptual speed and manual dexterity, was studied in 12 healthy male subjects. Each subject was tested during exposure to 250 ppm, 350 ppm, 450 ppm and 550 ppm of methylchloroform in inspiratory air and under control conditions with exposure to pure air. A linear relationship was noted between the concentration in alveolar air and arterial blood. Subject reaction time, perceptual speed and manual dexterity were all impaired during exposure. Analyses on the results strongly suggest that psychophysiological functions in humans are unfavorably affected by an exposure to an average concentration of 359 ppm (HSDB 1990).

Four groups of workers (total number = 196) exposed for at least 5 years to average concentrations of 4, 25, 28 and 53 ppm methylchloroform were evaluated for adverse effects. Routine laboratory examinations, including peripheral hemograms, blood-specific gravity and urinalysis, plus sense-of-vibration test were used. No consistent dose-related adverse effects were observed (HSDB 1990).

To determine whether unchanged solvent urinary concentration could be used as a biological exposure index, workers occupationally exposed to various solvents were studied. Nine unrelated groups working in plastic boat, chemical, plastic button, paint, and shoe factories were studied. A total 659 males were monitored. Urine samples were collected at the beginning of the workshift and at the end of the first half of the shift. A close relationship (correlation coefficient always above 0.85) between the average environmental solvent concentration measured in the breathing zone and the urinary concentration of unchanged solvent was observed. The proposed Biological Equivalent Exposure Limit (805 μg/L) corresponded to the environmental Threshold Limit Value (860 μg/L) for 1,1,1-trichloroethane. Biological exposure data for urine collected over 4 hours during random sampling for at least 1 year could be used to evaluate long-term exposure for individuals or groups of workers (HSDB 1990).

Human Carcinogenicity - The weight-of-evidence classification for 1,1,1-trichloroethane is "D; not classifiable as to human carcinogenicity." The basis for this classification is there are no reported human data and animal studies (one lifetime gavage, one intermediate-term inhalation) that have not demonstrated carcinogenicity. Technical-grade 1,1,1-trichloroethane has been shown to be weakly mutagenic, although the contaminant, 1,4-dioxane, a known animal carcinogen, may be responsible for this response (IRIS 1990).

Animal Carcinogenicity - The animal carcinogenicity data are considered inadequate. The NCI (1977), as cited in IRIS 1990, treated 50 male and 50 female Osborne-Mendal rats with 750 or 15000 mg/kg technical-grade 1,1,1-trichloroethane 5 times/wk for 78 weeks by gavage. The rats were observed for an additional 32 weeks. Twenty rats of each sex served as untreated controls. Low survival of both male and female treated rats (3%) may have precluded significant development of tumors late in life. Although a variety of neoplasms was observed in both treated and matched control rats, they were common to aged rats and were not significantly related to dosage. Similar results were obtained when the NCI (1977), as cited in IRIS 1990, treated B6C3F1 hybrid mice with the time-weighted average doses of 2807 or 5615 mg/kg 1,1,1,-trichloroethane by gavage 5 day/wk for 78 weeks. The mice were observed for an additional 12 weeks. The control and treated groups had 20 and 50 animals of each sex, respectively. Only 25% to 45% of those treated survived until the time of terminal sacrifice. A variety of neoplasms were observed in treated groups but with an incidence not statistically different from matched controls (IRIS 1990).

Quest et al. (1978), as cited in IRIS 1990, exposed 96 Sprauge-Dawley rats of both sexes to 875 or 1750 ppm 1,1,1'-trichloroethane vapor for 6 h/day, 5 day/wk for 12 months, followed by an additional 19-month observation period. The only significant sign of toxicity was an increased incidence of focal hepatocellular alterations in female rats at the highest dosage. It was not evident that a MTD was used nor was a range-finding study conducted. No significant dose-related neoplasms were reported, but these dose levels were below those used in the NCI study (IRIS 1990).

Carcinogenicity Supporting Data - Mutagenicity testing of 1,1,1-tri-chloroethane has produced positive results in S. typhimurium strain TA100 (Simmon et al. 1977; Fishbein 1979; Snow et al. 1979, all as cited in IRIS 1990) as well as some negative results (Henschler et al. 1977; Taylor 1978, as cited in IRIS 1990).

It was mutagenic for S. typhimurium strain TA1535 both with exogenous metabolic activation (Farber 1977, as cited in IRIS 1990) and without activation (Nestmann et al. 1980, as cited in IRIS 1990). 1,1,1-trichloro-ethane did not result in gene conversion or mitotic recombination in Saccharomyces cerevisiae (Farber 1977; Simmon et al. 1977, as cited in IRIS 1990) nor was it positive in a host-mediated forward mutation assay using Schizosaccharomyces pombe in mice. The chemical also failed to produce chromosomal aberrations in the bone marrow of cats (Rampy et al. 1977, as cited in IRIS 1990), but responded positively in a cell transformation test with rat embryo cells (Price et al. 1978, as cited in IRIS 1990).

An isomer, 1,1,2-trichloroethane, is carcinogenic in mice, inducing liver cancer and pheochromocytomas in both sexes. Dichloroethanes, tetra-chloroethane and hexachloroethane also produced liver cancer in mice and other types of neoplasms in rats. It should be noted that 1,4-dioxane, a known animal carcinogen, is a contaminant of technical-grade 1,1,1-trichloroethane. It causes liver and nasal tumors in more than one strain of rats and hepatocellular carcinomas in mice (IRIS 1990).

Quantitative Estimate of Carcinogenic Risk - No quantitative estimate of carcinogenic risk for either oral or inhalation exposure exists.

Toxicity Ranges - The acute lethal dose to the human has been estab-lished at 500 to 5,000 mg/kg. The 1,1,1-trichloroethane has a TLV in air of 350 ppm as compared to the 10 ppm TLV of 1,1,2-trichloroethane (HSDB 1990).

From the available data, it can be estimated that a single exposure to concentrations of methylchloroform of less than 5,000 ppm is probably not life-threatening to humans, though higher concentrations may produce central nervous system depression and possibly death (HSDB 1990).

A summary of the estimated dose-response relationships for acute effects of single short-term exposures is greater than 5000 ppm, which is the onset of central nervous system depression, and could be life threatening. A concentration of 2650-1900 ppm produces light-headedness and irritation of the throat. A concentration of 1000 ppm produces disturbance of equilibrium. Concentrations of 500 to 350 ppm produce slight changes in perception and obvious odor. A concentration of 100 ppm is the apparent odor threshold. It should be noted that these approximations are crude estimates and contain several areas of uncertainty (HSDB 1990).

2.2.21 Trichloroethylene

Trichloroethylene (TCE) is a mobile liquid with a characteristic odor of Chloroform. It has a density of 1.4649 and a flash point of 89.6°F. Its molecular formula is C_2HCl_3 and its molecular weight is 131.38. Trichloroethylene has a vapor pressure of 100 mm at 32°C. It has a Hazard Rating of 3 (Sax and Lewis 1989). The TLV, on a TWA basis, is 50 ppm. The TLV, on a STEL basis, is 200 ppm (ACGIH 1988). Some of the common synonyms for Trichloroethylene are acetylenetrichloride, benzinol, 1-chloro-2,2,-dichloroethylene, circosolv, 1,1-dichloro-2-chloroethylene, ethyinyltrichloride, ethylenetrichloride, lanadin, narcogen, TCE, TRI, triasol, trichlorethene, trichloroethene, 1,1,2-trichloroethylene, 1,2,2-trichloroethylene, tri-clene, triol, vitran, and westrosol.

2.2.21.1 Summary

Past health assessment information on TCE is under review by EPA. The past carcinogen assessment summary by EPA has been withdrawn and a new one is under development by the EPA CRAVE Work Group (IRIS 1990). A risk assessment for TCE is underreview by an EPA work group (IRIS 1990). Thus, past assessment information may be considerably modified in the future.

Trichloroethylene is a probable human carcinogen (EPA Group B2) and, according to EPA's preliminary risk assessment from ambient air exposures, public health risks are significant (4.1 cancer cases per year and maximum lifetime individual risks of 9.4E-5). Thus, EPA indicated that it will consider adding TCE to the list of hazardous air pollutants for which it will establish emission standards under section 112(b)(1)(A) of the Clean Air Act. The EPA will decide whether to add TCE to the list only after studying possible techniques that might be used to control emissions and further assessing the public health risks. The EPA will add TCE to the list if emissions standards are warranted (IRIS 1990).

Exposures to TCE vapors causes irritation of the mucous membranes with resultant conjunctivitis and rhinitis. Skin contact produces severe erythema and vesiculation followed by exfoliation. Ingestion causes a burning sensation in the mouth, nausea, vomiting and abdominal pain. Chronic exposure, as in solvent abusers, can produce weight loss, nausea, fatigue, visual impairment, dermatitis, and rarely jaundice (HSDB 1990).

Workers exposed at concentrations averaging about 10 ppm complained of headache, dizziness and sleepiness (HSDB 1990).

Prolonged occupational exposure to TCE has been associated with impairment of peripheral nervous system function, persistent neuritis and temporary loss of tactile sense, and paralysis of the fingers after direct contact with the solvent (HSDB 1990).

The Nuclear Regulatory Commission (NRC) in their Drinking Water and Health Volume 5 1983 report, as cited in HSDB (1990), estimated a human lifetime carcinogenic risk of 3.77E-7 for males and 6.84E-8 for females, assuming a daily consumption of 1 L of water containing trichloroethylene in a concentration of 1 μg/L.

2.2.21.2   Health Effects

Past health assessment information on TCE is under review by EPA. The past carcinogen assessment summary of EPA has been withdrawn and a new one is currently under development. Thus, a potential exists for considerable modification to past assessment information. The EPA published the Health Assessment Document for Trichloroethylene. Final Report in July 1985, EPA/600/8-82/006F (EPA 1985); and a Draft Addendum to the report, SAB-EHC-88-012, in March 1988 (EPA 1988). However, these reports were not used in this health effects assessment report, but are cited here for reference purposes. These EPA documents were not used because EPA is currently redoing their health assessment of trichloroethylene and will likely make considerable change to some of the positions reflected in these documents.

Noncarcinogenic Effects - A risk assessment for TCE is under review by an EPA work group (IRIS 1990).

Exposures to its vapors causes irritation of the mucous membranes with resultant conjunctivitis and rhinitis. Skin contact produces severe erythema and vesiculation followed by exfoliation. Ingestion causes a burning sensation in the mouth, nausea, vomiting and abdominal pain. Chronic exposure, as in solvent abusers, can produce weight loss, nausea, fatigue, visual impairment, dermatitis, and rarely jaundice (HSDB 1990).

Workers exposed at concentrations averaging about 10 ppm complained of headache, dizziness and sleepiness (HSDB 1990).

Prolonged occupational exposure to trichloroethylene has been associated with impairment of peripheral nervous system function, persistent neuritis and temporary loss of tactile sensem, and paralysis of the fingers after direct contact with the solvent (HSDB 1990).

Following chronic and acute overexposure of trichloroethylene during operation of a dry-cleaning unit, symptoms included symmetrical bilateral V)IIth cranial nerve deafness as well as cerebral cortical dysrhythmia and alterations in the electroencephalogram (EEG). The patient recovered after the exposure stopped (HSDB 1990).

The behavioral effects of exposure to TCE indicate that laboratory and work-place exposure to 540 or 1030 mg/m^3 for 70 minutes has no effect on reaction time or short-term memory (HSDB 1990).

Evoked trigeminal potentials were studies in 104 subjects occupationally exposed to TCE. Subjects had an average exposure time of 8.23 years and an average daily exposure of 7 hours (exposure levels were not given). Controls were 52 healthy nonexposed subjects. Symptoms suffered by 49 of the exposed subjects included dizziness, headache, asthenia, insomnia, mood perturbation, and sexual problems. Eighteen subjects had trigeminal nerve symptoms. These subjects were significantly older (P,0.001) than asymptomatic subjects. Forty subjects had a pathological trigeminal somatosensory evoked potential (TSEP). Of these, 28 had a normal trigeminal examination and 12 an abnormal one. For those with trigeminal symptoms, an abnormal TSEP was observed in subjects who had the longest and most intense exposure periods (HSDB 1990).

Human Carcinogenicity - The carcinogen assessment summary for TCE has been withdrawn following further review. A new carcinogen summary is in preparation by the EPA CRAVE Work Group (IRIS 1990).

Trichloroethylene is a probable human carcinogen (EPA Group B2) and, according to EPA's preliminary risk assessment from ambient air exposures, public health risks are significant (4.1 cancer cases per year and maximum lifetime individual risks of 9.4E-5). Thus, EPA indicated that it will consider adding TCE to the list of hazardous air pollutants for which it intends to establish emission standards under section 112(b)(1)(A) of the Clean Air Act. The EPA will decide whether to add TCE to the list only after studying possible techniques that might be used to control emissions and further assessing the public health risks. The EPA will add TCE to the list if emissions standards are warranted (IRIS 1990).

An epidemiology study on the hepatic tumor incidence in subjects working with TCE failed to show a correlation between liver cancer and occupational exposure. Another study, which looked at the mortality of 2117 workers exposed to TCE, found no correlation between cancer and occupational exposure (HSDB 1990).

Major consideration must be given to cumulative effects of trichloro-ethylene in long-term feeding studies carried out by the National Cancer Institute. Mice (both sexes, at both low and high dose levels) experienced a highly significant increase in hepatocellular carcinomas. Mikiskova and mikiska (1966), as cited in HSDB (1990), demonstrated that TCE had a pronounced depressant effect on the central nervous system (HSDB 1990).

The Carcinogen Assessment Group (CAG), Office of Health and Environmental Assessment in EPA's Research and Development Office, has prepared a list of chemical substances for which substantial or strong evidence exists showing that exposure to these chemicals, under certain conditions, causes cancer in humans, or can cause cancer in animal species that makes them potentially carcinogenic in humans. Substances are placed on the CAG list only if they have been demonstrated to 1) induce malignant tumors in one or more animal species, or 2) induce benign tumors that are generally recognized as early stages of malignancies, and/or positive epidemiologic studies indicated they were carcinogenic. Trichloroethylene is on the CAG list (HSDB 1990).

Quantitative Estimate of Carcinogen Risk - The NRC in their Drinking Water and Health Volume 5 1983 report, as cited in HSDB (1990), estimated human lifetime carcinogenic risk of 3.77E-7 for males and 6.84E-8 for females, assuming a daily consumption of 1 L of water containing TCE in a concentration of 1 μg/L.

Toxicity Ranges - The estimated dose in humans to cause death is reported to be 3 to 5 ml/kg. Eye irritation is reported at 160 ppm. The lowest concentration producing unconsciousness in adult humans is 16 mg/L; the equivalent oral dose is 40-150 ml (HSDB 1990)

Toxicant Interactions - Vasodilation and malaise ("degreasers' flush") occurs in workers who drink ethanol after exposure to Trichloroethylene (HSDB 1990).

2.2.22   1,2,4-Trimethylbenzene

1,2,4-trimethylbenzene is a liquid with a density of 0.888 and a flash point of 130°F. Its molecular formula is C_9H_{12} and its molecular weight is 120.21 (Sax and Lewis 1989). The TLV, on a TWA basis, is 25 ppm (ACGIH 1988). Some common synonyms are psi-cumene, pseudocumene, pseudocumol, 1,2,5-trimethylbenzene, and trimethylbenzene.

2.2.22.1   Summary

Trimethylbenzene is irritating to skin, eyes, and mucous membranes. It can cause central nervous system depression and thrombocytopenia. Asthmatic bronchitis may be provoked by exposure to trimethylbenzene. Chemical pneumonitis or pulmonary edema may develop, as well as headache, fatigue, nausea, and anxiety may be noted with exposures to trimethylbenzene.

Pertinent data regarding carcinogenicity of trimethylbenzenes to humans from oral or inhalation exposure could not be located by EPA in the available literature. Thus, trimethylbenzenes are best classified in EPA "Group D; Not classified" (EPA 1987).

There are no adequate subchronic or chronic inhalation or oral data that define dose-specific adverse effects. Therefore, no RfDs are available for trimethylbenzene.

The rat inhalation LC50 is 18 g/m^3/4 hours. The guinea pig intra-peritoneal LDLo is 1788 mg/kg; the rat intraperitoneal LDLo is 1752 mg/kg (RTECS 1987).

2.2.22.2 Health Effects

Noncarcinogenic Toxicity - Bernshtein (1972), as cited in EPA (1987), reported that phagocytic activity of leukocytes was inhibited in rats after inhalation of a mixture of trimethylbenzenes (1 mg/L, 1000 mg/m^3) 4 h/day, 6 day/wk for 6 months (EPA 1987).

Wiglusz et al. (1975a,b), as cited in EPA (1987), reported "slight" alteration in differential white blood cell count and elevated SGOT in male rats that had been exposed to 1,3,5-trimethylbenzene (3 mg/L, 3000 mg/m^3) 6 h/day, 6 day/wk for 5 weeks (EPA 1987).

Battig et al. (1957), as cited in EPA 1987, reported symptoms of nervousness, tension, anxiety and asthmatic bronchitis in a "significant number" of 27 people who worked for several years with a solvent containing 30% 1,3,5-Trimethylbenzene and 50% 1,2,4-trimethylbenzene. Tendencies toward hyperchromic anemia and blood coagulation were also observed among these individuals. Concentrations ranges for hydrocarbon vapor ranged from 10-60 ppm. Gerarde (1960), as cited in EPA (1987), speculated that a small proportion of benzene in the hydrocarbon vapor was probably responsible for the hematologic effects (EPA 1987).

In cases where paint thinner was used that contained more than 60% of mesitylene and pseudocumene, there were symptoms of disturbance of blood coagulation and a tendency to hematoma formation with low levels of thrombocytes and erythrocytes. Bronchitis, headache, fatigue and drowsiness was experienced by 70% of workers exposed to high concentrations (HSDB 1990).

Human Carcinogenicity - Pertinent data regarding carcinogenicity of trimethylbenzenes to humans from oral or inhalation exposure could not be located by EPA in the available literature. Thus, trimethylbenzenes are best classified in EPA "Group D; Not classified" (EPA 1987).

IARC has not evaluated the weight of evidence for carcinogenicity to humans of trimethylbenzenes. Because data are inadequate, an IRAC classification of Group 3 is most appropriate. According to the EPA guidelines for evaluating the weight of evidence, and EPA classification of Group D (not classified) best reflects the lack of carcinogenicity test data (EPA 1987).

Quantitative Estimate of Carcinogen Risk - There are no adequate subchronic or chronic inhalation or oral data that define dose-specific adverse effects. Therefore, no RfDs are available for trimethylbenzene.

Animal Toxicity - The rat inhalation LC50 is 18 g/m^3/4 hours. The guinea pig intraperitoneal LDLo is 1788 mg/kg, the rat intraperitoneal LDLo is 1752 mg/kg (RTECS 1987).

Toxicant Interactions - Benzene and its methyl derivatives are metabolized to derivatives of phenol and hippurate. Oral administration of benzene along with either 1,2,3- or 1,3,5-trimethylbenzene resulted in a higher concentration of phenol in the blood than when benzene was administered alone. Hippuric acid levels in the blood were also elevated when benzene was administered together with 1,2,3-trichlorobenzene, but were decreased when benzene was administered with 1,3,5-trimethylbenzene (Mikulski et al. 1979, as cited in EPA 1987).

2.2.23 Undecane

Undecane is a colorless liquid with a density of 0.7402 and a flash point of 149°F. Its molecular formula is $C_{11}H_{24}$ and its molecular weight is 156.35. It has a Hazard Rating of 2 (Sax and Lewis 1989). Common synonyms for undecane are hendecane and n-undecane.

2.2.23.1 Summary

The amount of information regarding toxicological effects of undecane is very limited. No studies of human toxicology were found.

The only health effect related limit found was the intravenous LD50 in mice of 517 mg/kg.

2.2.23.2 Health Effects

The amount of information regarding toxicological effects of undecane is very limited. No studies of human toxicology were found.

General Toxicity - The direct aspiration into the lungs of paraffins with carbon numbers C_6 to C_{16} may cause chemical pneumonitis, pulmonary edema and hemorrahaging (HSDB 1990).

An intravenous LD50 in mouse is 517 mg/kg, with a statement that the toxic effects have not yet been reviewed (RTECS 1987).

Sax and Lewis (1989) make a statement that undecane is moderately toxic by the intravenous route, and they give undecane a Hazard Rating of 2.

2.2.24 Xylene

Xylene is a clear liquid with a density of 0.864 and a flash point of 100°F. It has a molecular formula of C_8H_{10} and it has a molecular weight of 106.18. Xylene has a vapor pressure of 6.72 mm at 21°C. It has a Hazard Rating of 3 (Sax and Lewis 1989). The TLV, on a TWA basis, is 100 ppm. The TLV, on a STEL basis, is 150 ppm. These TLVs are for the o-, m-, and p-isomers of xylene (ACGIH 1988). Common synonyms for xylene are dimethyl-benzene, 1,2-dimethylbenzene, 1,3-dimethylbenzene, 1,4-dimethylbenzene, m-xylene, meta-xylene, o-xylene, ortho-xylene, p-xylene, and para-xylene.

2.2.24.1 Summary

The RfDo for xylene is 2E+0 mg/kg/day, with the critical effect being hyperactivity, and decreased body weight. The NOAEL for xylene is 250 mg/kg/day, and the FEL is 500 mg/kg/day (IRIS 1990).

Xylene is a central nervous system depressant that produces lightheaded-ness, nausea, headache, and ataxia at low doses and confusion, respiratory

depression, and coma at high doses. Above 200 ppm, xylene causes conjunctivitis, nasal irritation, and sore throats. It is a potent respiratory
irritant at high concentrations (HSDB 1990).

Central nervous system defects were more common in children of mothers
exposed to organic solvents and dusts during pregnancy. Hydranencephaly
occurred in children whose mothers had been exposed to the solvents toluene,
xylene, and White Spirit during manufacture of rubber products (HSDB 1990).

Women exposed to xylene are liable to suffer from menstrual disorders
(menorrhagia, metrorrhagia). It has been reported that female workers exposed
to xylene in concentrations that periodically exceed the exposure limits were
also affected by pathological pregnancy conditions (toxicosis, danger of mis-
carriage, hemorrhage during childbirth) and infertility (HSDB 1990).

The weight-of-evidence classification for xylene is "D; not classifiable
as to human carcinogenicity." The basis for this classification is the orally
administered technical xylene mixtures did not result in significant increases
in incidences in tumor responses in rats or mice of both sexes (IRIS 1990).

2.2.24.2 Health Effects

Noncarcinogenic Effects - The RfD is based on the assumption that
thresholds exist for certain toxic effects such as cellular necrosis, but may
not exist for other toxic effects such as carcinogenicity. In general, the
RfD is an estimate of a daily exposure to the human population (including
sensitive subgroups) that is likely to be without an appreciable risk of
deleterious effects during a lifetime. The RfDo for xylene is 2E+0 mg/kg/day,
with the critical effect being hyperactivity, and decreased body weight. The
NOAEL for xylene is 250 mg/kg/day, and the FEL is 500 mg/kg/day (IRIS 1990).

Groups of 50 male and 50 female Fischer 344 rats and 50 male and
50 female B6C3F1 mice were given gavage doses of 0, 250, or 500 mg/kg/day
(rats) and 0, 500, or 1000 mg/kg/day (mice) for 5 day/wk for 103 weeks. The
animals were observed for clinical signs of toxicity, body weight gain, and
mortality. All animals that died or were killed at sacrifice were given gross
necropsy and comprehensive histological examinations. There was a dose-
related increased mortality in male rats, and the increase was significantly

greater in the high-dose group compared with controls. Although increased mortality was observed at 250 mg/kg/day, the increase was not significant. Although many of the early deaths were caused by gavage error, NTP (1986), as cited in IRIS (1990), did not rule out the possibility that the rats were resisting gavage dosing because of the behavioral effects of xylene. Mice given the high dose exhibited hyperactivity, a manifestation of central nervous system toxicity. There were no compound-related histopathologic lesions in any of the treated rats or mice. Therefore, the high dose is a FEL and the low dose a NOAEL (IRIS 1990).

EPA (1984), as cited in IRIS (1990), reported an RfD of 0.01 mg/kg/day, based on a rat dietary NOAEL of 200 ppm or 10 mg/kg/day as defined by Bowers er al. (1982), as cited in IRIS (1990), in a 6-month study. This NOAEL was divided by an uncertainty factor of 1,000. EPA (1985, 1986), as cited in IRIS (1990), noted that this study used aged rats, loss of xylene from volatilization was not controlled, only one exposure level was used, and histopathologic examination was incomplete. An RfD of 4.31 mg/day (about 0.06 mg/kg/day) based on an inhalation study (Jenkins et al. 1970, as cited in IRIS 1990) using rats, guinea pigs, monkeys, and dogs exposed to o-Xylene at 3358 mg/m^3, 8 h/day, 5 day/wk for 6 weeks or at 337 mg/m^3 continuously for 90 days was derived by DOE (1985), as cited in IRIS (1990). Deaths in rats and monkeys, and tremors in dogs, occurred at the highest dose, whereas no effects were observed in the 337 mg/m^3 continuous exposure group. The RfD based on the NTP study (1986), as cited in IRIS (1990), is preferable because it is based on a chronic exposure in two species by a relevant route of administration, and comprehensive histology was performed. Xylene is fetotoxic and teratogenic in mice at high oral doses (Nawrot and Staples 1981; Marks et al. 1982, as cited in IRIS 1990), but the RfD as calculated should be protective of these effects (IRIS 1990).

The confidence in the RfDo for xylene is medium. The NTP study (1986), as cited in IRIS (1990), was given a medium confidence level because it was a well-designed study in which adequately sized groups of two species were tested over a substantial portion of their lifespan, comprehensive histology was performed, and a NOAEL was defined; but clinical chemistries, blood

enzymes, and urinalysis were not performed. The database was given a medium
confidence level because, although supporting data exist for mice and
teratogenicity and fetotoxicity data are available with positive results at
high oral doses, a LOAEL for chronic oral exposures has not been defined. The
medium confidence in the RfDo follows (IRIS 1990).

Xylene is a central nervous system depressant that produces lightheaded-
ness, nausea, headache, and ataxia at low doses and confusion, respiratory
depression, and coma at high doses. Above 200 ppm, xylene causes conjunctivitis, nasal irritation, and sore throats. It is a potent respiratory
irritant at high concentrations (HSDB 1990).

Transient mildly elevated hepatic aminotransferase levels and reversible
renal failure were reported in an estimated 10,000 ppm xylene exposure occurring during the painting of a poorly ventilated ship compartment. Two men
were comatose and one was dead on arrival after prolonged exposure over
18 hours. The survivors developed no long-term sequelae. The contributions
of hypoxia and a toluene solvent could not be quantitated (HSDB 1990).

Six volunteers were able to detect odor of mixed xylenes at concentrations of 60 mg/m^3 in a 15-minute exposure period. The only common sign of
discomfort at 2,000 mg/m^3 was eye irritation. Some olfactory fatigue occurred
with recovery in 10 minutes (HSDB 1990).

In workers exposed to organic solvents (acetone, benzene, toluene, ethyl
acetate, butyl acetate, xylene, gasoline, and turpentine), the incidence of
chronic bronchitis was higher and the volume of expiratory air was lower than
in normal control subjects. In smokers, the incidence was higher than nonsmokers of exposed and nonexposed groups. Smoking increases risk of chronic
bronchitis in subjects exposed to organic solvents (HSDB 1990).

Central nervous system defects were more common in children of mothers
exposed to organic solvents and dusts during pregnancy. Hydranencephaly
occurred in children whose mothers had been exposed to the solvents toluene,
xylene, and White Spirit during manufacture of rubber products (HSDB 1990).

Exposed women are liable to suffer from menstrual disorders (menorrhagia, metrorrhagia). It has been reported that female workers exposed to

xylene in concentrations that periodically exceed the exposure limits were
also affected by pathological pregnancy conditions (toxicosis, danger of mis-
carriage, hemorrhage during child birth) and infertility (HSDB 1990).

Human Carcinogenicity - The weight-of-evidence classification for xylene
is "D; not classifiable as to human carcinogenicity." The basis for this
classification is the orally administered technical xylene mixtures did not
result in significant increases in incidences in tumor responses in rats or
mice of both sexes (IRIS 1990).

Animal Carcinogenicity - The animal carcinogenicity data are considered
inadequate. In an NTP study (1986), as cited in IRIS (1990), 50 male and
50 female F344/n rats were treated by gavage with mixed xylenes in corn oil
(60% m-xylene, 14% p-xylene, 9% o-xylene and 17% ethylbenzene) at dosages of
0, 250 or 1000 mg/kg/day. Animals were killed and examined histologically
when moribund or after 104-105 weeks. An apparent dose-related increased
mortality was observed in male rats, but this difference was statistically
significant for the high dose group only. No other differences in survival
between dosage groups of either sex were observed. Interstitial cell tumors
of the testes could not be attributed to administration of the test compound
observed in male rats (43/50 control, 38/50 low-dose and 41/40 high-dose).
NTP (1986), as cited in IRIS (1990), reported no significant changes in the
incidence of neoplastic or nonneoplastic lesions in the rats or mice that
could be considered related to the mixed Xylene treatment, and concluded that
under conditions of these 2-year gavage studies, there was "no evidence of
carcinogenicity" of xylene (mixed) for rats or mice of either sex at any
dosage tested (IRIS 1990).

Maltoni et al. (1985), as cited in IRIS (1990), in a limited study,
reported higher incidences (compared with controls) of malignant tumors in
male and female Sprague-Dawley rats treated by gavage with xylene in olive oil
at 500 mg/kg/day, 4 or 5 day/wk for 104 weeks. This study did not report
survival rates or specific tumor types; therefore, the results cannot be
interpreted (IRIS 1990).

Berenblum (1941), as cited in IRIS (1990), reported that undiluted
xylene applied at weekly intervals produced one tumor-bearing animal out of

40 after 25 weeks in skin-painting experiments in mice. No control groups
were described. Pound (1970), as cited in IRIS (1990), reported negative
results in initiation-promotion experiments with xylene as the initiator and
cronton oil as the promotor.

Carcinogenicity Supporting Data - The frequency of sister chromatid
exchanges and chromosomal aberrations were nearly identical between a group of
17 plant industry workers exposed to xylene and their respective referents
(Haglund et al. 1980, as cited in IRIS 1990). In vitro, xylene caused no
increase in the number of sister chromatid exchanges in human lymphocytes
(Gerner-Smidt and Friedrich 1978, as cited in IRIS 1990). Studies indicate
that xylene isomers, technical grade xylene or mixed xylene are not mutagenic
in tests with Salmonella typhimurium (Florin et al. 1980; NTP 1986; Bos et al.
1981, all as cited in IRIS 1990) nor in mutant reversion assays with
Escherichia coli (McCarrol et al. 1981, as cited in IRIS 1990). Technical
grade xylene, but not o- and m-xylene, was weakly mutagenic in Drosophila
recessive lethal tests. Chromosomal aberrations were not increased in bone
marrow cells of rats exposed to xylenes by inhalation (Donner et al. 1980, as
cited in IRIS 1990).

Quantitative Estimate of Carcinogen Risk - There are no quantitative
estimates of risk for either the oral or the inhalation routes.

2.2.25 Additional Calk-Associated Compounds

The following three compounds are additional contaminants found in
caulking material. In the reviewed public building measurement studies, these
contaminants were not specifically reported as being found, but they are
included here for completeness. It is suspected that if these contaminants
are detected in building caulking material, they are very likely to be
released to the indoor air of a public building.

2.2.25.1 Carbitol

Carbitol is a colorless liquid with a mild pleasant odor. It has a
density of 0.9902 and a flash point of 201°F. Its molecular formula is
$C_6H_{14}O_3$ and its molecular weight is 134.2. It has a Hazard Rating of 2 (Sax
and Lewis 1989). The most common synonym for carbitol is carbitol cellosolve,

with carbitol solvent, diethylene glycol ethyl ether, dioxitol, diglycol monoethyl ether, ethoxy diglycol, 2-(2-ethoxyethoxy)ethanol, ethylcarbitol, ethyldiethyleneglycol, poly-solv, and solvsol other common synonyms.

2.2.25.1.1 Summary. The amount of information regarding toxicological effects of carbitol is very limited. No studies of human toxicology were found.

2.2.25.1.2 Health Effects. The amount of information regarding toxicological effects of carbitol is very limited. No studies of human toxicology were found.

General Toxicity - Carbitol is moderately toxic by ingestion, intravenous, intraperitoneal and possibly other routes. It is mildly toxic by skin contact, and it is a skin and eye irritant. Experimental reproductive effects have been associated with it (Sax and Lewis 1989).

Skin and eye irritation levels are shown in RTECS at 112 mg/3 days with mild effects for human skin. A rabbit skin level of 500 mg with mild irritation is provided. A rabbit eye irritation level of 50 mg with mild irritation is listed (RTECS 1987).

A reproductive effects level for rats, administered orally, is provided showing an LDLo of 50 mg/kg resulting in specific developmental abnormalities (musculoskeletal system) (RTECS 1987).

Three values of tumorigenic data for rats, administered orally, are provided. These levels are 1) oral rat TDLo of 890 gm/kg/53 weeks, 2) oral rat TD of 1752 gm/kg/2 years, and 3) oral rat TD of 584 gm/kg/2 years. The tumors developed were in the kidney, ureter and bladder (RTECS 1987).

Toxicity data are also provided in RTECS. A mouse inhalation LCLo of 130 mg/m^3/2 hours with toxic effects not yet reviewed is provided. A rat intramuscular LDLo value of 7826 mg/kg with no toxic effects is listed. A rabbit intramuscular LDLo value of 4472 mg/kg with no toxic effects noted is listed. The following is provided: a mouse intraperitoneal LD50 of 9719 mg/kg with lungs, thorax or respiration (chronic pulmonary edema or congestion); kidney, ureter and bladder (changes in both tubules and glomeruli); and blood (changes in spleen) affected. A rat intravenous LD50 of 6565 mg/kg with toxic effects not yet reviewed is listed. Also, a rabbit intravenous LDLo of 2236 mg/kg with no toxic effects noted is listed. A mouse subcutaneous LDLo of 5 gm/kg with toxic effects not yet reviewed is listed. A rat subcutaneous LDLo of 16770 mg/kg with no toxic effects noted is listed. A rabbit skin LD50 of 11890 mg/kg with no toxic effects noted is provided (RTECS 1987).

2.2.25.2 Ethylene Glycol

Ethylene glycol is a colorless, sweet-tasting, hygroscopic liquid. It has a density of 1.113 and a flash point of 232°F. Its molecular formula is $C_2H_6O_2$ and its molecular weight is 62.08. Ethylene glycol has a vapor pressure of 0.05 mm at 20°C (Sax and Lewis 1989). The TLV, on a TWA basis, is 50 ppm (ACGIH 1988). Common synonyms for ethylene glycol are 1,2-dihydroxyethane, 1,2-ethanediol, ethylene alcohol, ethylene dihydrate, glycol, glycol alcohol, MEG, monoethylene glycol, norkool, and tescol.

2.2.25.2.1 Summary. The amount of toxicological information available for ethylene glycol is somewhat limited.

The RfDo for ethylene glycol is 2E+0 mg/kg/day, with kidney toxicity listed as the critical effect. The NOAEL is 200 mg/kg/day, and the LOAEL is 1,000 mg/kg/day (IRIS 1990).

The RfDi is not available at this time.

Ethylene glycol is in the same range of toxicity as methylethylketone, propylenedichloride, and TCE, but its hazards are believed to be less because its vapor pressure is substantially lower (HSDB 1990).

Ethylene glycol has not been evaluated by the EPA for evidence of human carcinogenic potential.

2.2.25.2.2 Health Effects.

Noncarcinogenic Effects - The RfD is based on the assumption that thresholds exist for certain toxic effects, such as cellular necrosis, but may not exist for other toxic effects, such as carcinogenicity. In general, the RfD is an estimate of a daily exposure to the human population (including sensitive subgroups) that is likely to be without an appreciable risk of deleterious effects during a lifetime. The RfDo for ethylene glycol is 2E+0

mg/kg/day, with kidney toxicity listed as the critical effect. The NOAEL is
200 mg/kg/day, and the LOAEL is 1,000 mg/kg/day (IRIS 1990).

DePass et al. (1986a); as cited in IRIS (1990), conducted 2-year studies
using groups of approximately 30 rats per sex and 20 mice per sex that were
fed diets providing ethylene glycol dosages of 0, 40, 200, or 1,000 mg/kg/day.
High-dose rats had increased mortality, neutrophil count, water intake, kidney
hemoglobin and hematocrit, and chronic nephrities. Female rats exposed to
1,000 mg/kg/day had mild fatty changes in the liver. No adverse effects
occurred at other doses in rats or at any dose in mice (IRIS 1990).

Groups of Sprauge-Dawley rats (16/sex/group) were fed diets containing
0, 0.1, 0.2, 0.5, 1.0, or 4.0% ethylene glycol for 2 years (Bolld 1965, as
cited in IRIS 1990). Male rats at 1.0 and 4.0% and females at 4.0% had
increased mortality, decreased growth, increased water consumption,
proteinuria, and renal calculi. There was an increased incidence of
cytoplasmic crystal deposition in renal tubular epithelium at 0.5 and 1.0%.
There were no effects on organ weights or hematologic parameters. The authors
concluded that 0.2% (2,000 ppm) was a NOAEL for rats; the LOAEL was 0.5%
(5,000 ppm). Assuming that a rat consumes food equivalent to 5% of its body
weight/day, the NOAEL and the LOAEL are equivalent to 100 mg/kg/day and 250
mg/kg/day, respectively (IRIS 1990).

The choice of the DePass et al. study (1986a), as cited in IRIS (1990),
over the Blood study (1965), as cited in IRIS (1990), as the basis of the Rfd
reflects the greater confidence in the former study because of the greater
number of animals tested and effects considered. The magnitude of the RfD in
either case is similar (IRIS 1990).

Bolld et al. (1962), as cited in IRIS (1990), fed a diet containing 0.2
or 0.5% ethylene glycol to rhesus monkeys for 3 years. No treatment-related
toxic effects on histologic appearance of kidneys or other major organs were
found.

In a teratogenicity study, Maronpot et al. (1983), as cited in IRIS
(1990), found increased preimplantation loss and increased incidence of poorly
ossified vertebral centra in offspring of rats treated at 1,000 mg/kg in the

diet on days 6-15 of gestation. No effects occurred at 40 or 200 mg/kg. Lamb et al. (1985), as cited in IRIS (1990), reported that exposure of male and female mice to 1.0% ethylene glycol in drinking water for 14 weeks resulted in significantly fewer litters, decreased mean live pup weight, and decreased number of live pups per litter. DePas et al. (1986a), as cited in IRIS (1990), conducted a three-generation study in which rats were treated with 0, 40, 200, or 1,000 mg/kg/day and mice with 0, 750, 1500, or 3000 mg/kg/day on days 6-15 of gestation. The percentage of litters with malformed fetuses increased in a dose-related manner in both species at all doses. There was a dose-related increase in postimplantation losses per litter in both species, but it was significant only in high-dose rats. Maternal body weight gain was decreased at all doses in rats and at the two higher doses in mice (IRIS 1990).

The confidence in the RfDo is high. Confidence in the DePass et al. study (1986a), as cited in IRIS (1990), is rated high because it was a well-conducted lifetime study in two species by a relevant route and defined a NOAEL and LOAEL. Confidence in the database is also high because it contains another chronic rat study and a monkey study that support the NOAEL and LOAEL from the DePass et al. study (1986a), as cited in IRIS (1990). It also contains data that indicate that the RfDo is protective of teratogenic and reproductive effects. Therefore, confidence in the RfDo is also high (IRIS 1990).

The RfDi is not available at this time.

Ethylene glycol is in the same range of toxicity as methylethylketone, propylenedichloride, and TCE, but its hazards are believed to be less because its vapor pressure is substantially lower (HSDB 1990).

Human Carcinogenicity - Ethylene glycol has not been evaluated by the EPA for evidence of human carcinogenic potential.

2.2.25.3 <u>Octene</u>

Octene was not found in Sax and Lewis (1989), IRIS (1990), HSDB (1990), RTECS (1987), or in the various bibliographic databases searched. Thus, it was concluded that no toxicological information is available on octene at this time.

2.3 COMBUSTION PRODUCTS

Pollutant emissions from indoor combustion vary considerably as a function of the type of fuel, the type and condition of the appliance, and the conditions under which combustion takes place. The major combustion products are CO_2 and water vapor. Major combustion by-products are CO, nitrogen dioxide, sulfur dioxide, aldehydes and other organic compounds, and particulate matter. Tobacco smoking, gas-fired appliances, wood stoves, fireplaces, and kerosene heaters are some of the major sources of indoor-air contaminating combustion by-products.

Metabolic activity is responsible for most of the indoor CO_2 concentration in the indoor environment. Gas and kerosene heaters also emit appreciable quantities of CO_2. Carbon monoxide is a product of incomplete combustion. Emissions from vehicles in attached or underground garages or from outside traffic, with emissions being brought in through the ventilation systems, also contribute indoor air contaminants. Oxides of nitrogen are formed when combustion takes place at high temperatures. Sulfur oxides are combustion products of sources burning sulfur containing fuels (e.g., kerosene and coal). In general, sulfur oxides are usually more of a concern with outdoor ambient air quality.

Among the aldehydes and other organic compounds associated with combustion products, formaldehyde and benzo(a)pyrene are the sources of greatest concern. Formaldehyde was discussed earlier in Section 2.2.11. Wood combustion and tobacco smoke are the main sources of the normally fairly low levels of benzo(a)pyrene found in the indoor environment.

2.3.1 <u>Benzo(a)pyrene</u>

Benzo(a)pyrene [B(a)P] in its pure form is yellow crystals that are insoluble in water but soluble in benzene, toluene, and xylene. The American Conference of Governmental Industrial Hygienists do not list any TLVs for B(a)P. Instead, they note that it is a "Suspected human carcinogen and exposures to carcinogens must be kept to a minimum." B(a)P has a Hazard Ratng of 3 (Sax and Lewis 1989).

2.3.1.1 <u>Summary</u>

The weight-of-evidence classification for B(a)P is "B2; probable human carcinogen." Human data specifically linking B(a)P to a carcinogenic effect are lacking. However, there are multiple animal studies in rodent and non-rodent species demonstrating B(a)P to be carcinogenic following administration by oral, intratracheal, inhalation and dermal routes. B(a)P has produced positive results in several vitro bacterial and mammalian genetic toxicology assays (IRIS 1990).

The carcinogenicity observed in association with B(a)P is believed to come from its metabolic derivatives rather than the compound itself. The review of data on B(a)P as a pure compound shows that those derivatives can range from noncarcinogenic to extremely carcinogenic. Not only does carcinogenic response vary from metabolite to metabolite, but the response to B(a)P and/or its metabolites may vary from one species to another, one variety to another, and even among individuals within a variety. In light of this information, the use of an single dose-response relationship in a model for predicting cancer risk, even if it is based on extensive comparisons of actual populations of smoking and nonsmoking humans, is open to question (DOE 1987).

Recent animal studies on the carcinogenicity of complex mixtures such as coal liquids have demonstrated that the carcinogenicity of the mixture does not vary directly with its B(a)P content (DOE 1987).

B(a)P is a ubiquitous poly aromatic hydrocarbon (PAH) produced during the combustion of materials such as coal, wood, tobacco, diesel fuel, and tar, and in some commercial processes. It has been extensively studied and shown to be a potent carcinogen in experimental animals and in human tissues. As a

PAH, B(a)P is not chemically reactive with cellular components of living organisms. Rather, it is a procarciongen that is metabolized by living cells into carcinogenic derivatives capable of reacting directly with the macromolecules of animal and human cells (Kramer et al. 1983, as cited in DOE 1987). B(a)P has been widely used as a reference compound for estimating human health effects for emissions and effluent suspected of posing a carcinogenic risk to human populations (DOE 1987).

2.3.1.2   Health Effects - Pure B(a)P

Human Carcinogenicity - The weight-of-evidence classification for B(a)P is "B2; probable human carcinogen." Human data specifically linking B(a)P to a carcinogenic effect are lacking. However, there are multiple animal studies in rodent and nonrodent species demonstrating B(a)P to be carcinogenic following administration by oral, intratracheal, inhalation and dermal routes. B(a)P has produced positive results in several vitro bacterial and mammalian genetic toxicology assays (IRIS 1990).

Human carcinogenicity data are inadequate. Lung cancer has been shown to be induced in humans by various mixtures of polycyclic aromatic hydrocarbons known to contain B(a)P including cigarette smoke, roofing tar and coke oven emissions. It is not possible, however, to conclude from this information that B(a)P is the responsible agent (IRIS 1990).

No dose-response information on the experimental or occupational exposure of humans to pure B(a)P appears to exist. However, there have been a number of studies on the exposure of human organs, cells, and microsomal fractions cultured in vitro. The most significant finding of these studies has been that comparisons of limited human data with the more extensive animal data suggest basically similar effects of B(a)P (Kramer et al. 1983, as cited in DOE 1987). At the same time, differences in the reactions of different human organs have been demonstrated. For example, the urinary bladder appears to be most sensitive, followed by the skin, bronchus, esophagus, and colon. Moreover, there are significant variations in the susceptibility of tissues taken from different individuals, some of which appear to be associated with genetic differences (DOE 1987).

B(a)P is well known as a complete carcinogen when applied to the skin of mice, rats and rabbits (IARC 1973, as cited in IRIS 1990). Suscutaneous or intramuscular B(a)P injection has been shown to result in local tumors in mice, rats, guinea pigs, monkeys and hamsters (IARC 1973, as cited in IRIS 1990). Intratracheal instillation of B(a)P produced increased incidences of respiratory tract neoplasms in both male and female Syrian hamsters (Feron et al. 1973; Kobayashi 1975, as cited in IRIS 1990).

B(a)P administered orally to rats and hamsters produces stomach tumors. Neal and Rigdon (1967), as cited in IRIS (1990), administered dietary B(a)P at concentrations of 0, 1 10, 20, 30, 40, 45, 50, 100, and 250 ppm to male and female CFW-Swiss mice. The control group numbered 289; treatment groups varied in number from 9 to 73 animals and treatment time from 1 to 197 days. Stomach tumors were observed in mice consuming 20 or more ppm B(a)P. Incidence was apparently related both to the dose amount and the number of administered doses. Apparent increased incidences of leukemia and lung adenomas were reported in the mice on high B(a)P diets (250 and 1000 ppm) (Ridgon and Neal 1966; 1969, as cited in Iris 1990.

Tyssen et al. (1981), as cited in IRIS (1990), exposed groups of 24 hamsters by inhalation of B(a)P at concentrations of 2.2, 9.5, or 45 mg/m^3 for 4.5 h/day for 10 weeks followed by 3 h/day (7 day/wk) for up to 675 days. No animals in the lowest treatment group developed respiratory tumors. Those hamsters exposed to 9.5 mg/m^3 developed tumors of the nasal cavity, larynx, trachea, and pharynx. In addition to respirator tract tumors, animals in the highest dose group were seen to have neoplasms of the upper digestive tract (IRIS 1990).

B(a)P is among the best-studied agents producing genetic toxicological effects. It is metabolized to reactive electrophiles capable of binding to DNA. In vitro assays in which B(a)P has produced positive results include the following: bacterial DNA repair, bacteriophage induction, point mutations at multiple loci in several bacterial species and strains, mutations in Drosophila melanogaster, sister-chromatid-exchange, chromosomal aberrations and mutation and transformation of cultured mammalian cells. In vivo exposure of mammalian species to B(a)P has produced the following results: sister-chromatid-exchange, chromosomal aberrations, sperm abnormalities, and positive results in the mouse specific locus (spot) test (IARC 1973; 1983; Santodonato et al. 1981, all as cited in IRIS 1990).

Quantitative Estimate of Carcinogen Risk - No quantitative estimate of carcinogenic risk from oral or inhalation exposure is available at this time.

2.3.1.3   Health Effects - B(a)P in Complex Mixtures

Although studies of the effects of B(a)P on animals and on human tissues are useful in demonstrating the metabolic derivatives responsible for mutagenic/carcinogenic activity, or for determining differences between animals and humans, or among humans to different exposure scenarios, they have serious limitations in extrapolating to the environmental exposure of humans. One important limitation is that B(a)P is not normally encountered in the environment (either ambient or indoor) as a pure compound. Rather, it occurs with many other organics in what are commonly referred to as complex mixtures. For example, B(a)P is encountered in cigarette smoke along with about 1000 other compounds and in coal-derived products with as many as 2500 other compounds (DOE 1987).

Risk assessments performed in the 1970s and early 1980s tended to esti-mate the carcinogenic potential of these mixtures on the basis of their B(a)P content (Mahlum et al. 1984, as cited in DOE 1987). This practice continues. In a recent international symposium, it was concluded that for each individual type of polycyclic organic matter (POM) source, "both Benzo (A) Pyrene and total POM provide excellent indicators of hazard..." (Milliken et al. 1984, as cited in DOE 1987). As a result of this emphasis, many people regard B(a)P as the determinant carcinogen in many complex mixture pollutants, whether or not there is experimental evidence on which to base this assumption (DOE 1987).

Human Health Effect Studies - The evidence most cited for the carcino-genicity of B(a)P in humans comes from epidemiological studies of workers who have been occupationally exposed to this substance as a component of complex mixtures encountered. For example, EPA developed estimates of cancer risk for worker populations exposed to coke-oven emissions (EPA 1981, updated in EPA

1984b, as cited in DOE 1987). The EPA scientists developed a unit risk factor (the lifetime probability of dying from lung cancer as a result of exposure to coke-oven emissions) based on the benzene-soluble organic fraction of emissions, of which B(a)P is a component. Based on the rate of lung cancer among coke-oven workers, and assuming that there is no level of exposure that does not have some risk associated with it, the estimated lifetime risk of dying from lung cancer from continuous exposure to 1 μg of benzene-soluble organic per cubic meter of air has been estimated at 9.25E-4 (DOE 1987).

More commonly, humans are exposed to B(a)P as a component of cigarette smoke (NRC 1981, as cited in DOE 1987). Estimates of risk from B(a)P in the indoor environment given in The Expanded Weatherization Final Environmental Impact Statement (DOE 1984) are based on a model developed from the risk of lung cancer among cigarette smokers and tested using data from occupational populations exposed to coke-oven emissions. Again, the tendency is to attribute most or all of the carcinogenic activity of a complex mixture, this time cigarette smoke, to its B(a)P content. Evidence exists from both human and animal studies that this assumption is not reliable (DOE 1987).

A pilot study in four Indian villages of women exposed to total suspended particulates (TSP) and particulate B(a)P from cooking on simple stoves has been reported by Smith et al. (1983), as cited in DOE (1987). Stove fuels were traditional biomass fuels that resulted in TSP exposures averaging 7 mg/m^3 and approximately 4000 ng/m^3 B(a)P during the cooking period. A comparison of resulting exposure levels with smokers revealed that village cooks received more than a factor of 10 greater nominal doses to B(a)P. Yet, these women did not show incidences of lung or other cancers proportional to their B(a)P exposure (DOE 1987).

<u>Nonhuman Studies</u> - Pure B(a)P adsorbed to fine particles has a longer lung retention time (1 day for 90% clearance) than a pure B(a)P aerosol (4 hours for 90% clearance). A further increase in lung clearance time to about 60 days was shown in experiments with B(a)P plus diesel particles (complex mixture) (Sun et al. 1982, as cited in DOE 1987). Therefore, B(a)P associated with complex mixtures is retained in the lungs longer than pure B(a)P, whether bound to a fine particle or as an aerosol (DOE 1987).

Several studies conducted at the Pacific Northwest Laboratory (PNL) with coal liquids have suggested that the mutagenic and carcinogenic activities of B(a)P were inhibited when the B(a)P was contained in a complex mixture. Pelroy (1984), as cited in DOE 1987, using the Ames _Salmonella_ histidine reversion assay (used as a screening test for potential carcinogenicity) showed that the mutagenic activities of the neutral PAH fractions of high boiling coal distillates were very low relative to their B(a)P content. Moreover, the addition of small amount of coal distillate to the assay system substantially inhibited the mutagenic activity of exogenous B(a)P (DOE 1987).

Studies using the mouse skin tumor initiaton-promotion assay were performed to directly determine the carcinogenicity of a number of coal-derived liquids. From these studies, we have obtained numerous lines of evidence that the carcinogenic activity of B(a)P is substantially reduced when administered in a complex mixture (DOE 1987).

One line of evidence came from comparing the carcinogenicity of coal liquids from two different coal liquefraction processes, the heavy distillate (HD) from the solvent refined coal-II (SRC-II) process, and the finer feed from the Integrated Two-Stage Liquefaction (ITSL) process. Although the HD contained 300 ppm B(a)P and the finer feed 30,000 ppm B(a)P, the carcinogenic potencies of the two liquids were almost identical (DOE 1987).

A second line of evidence was obtained when the carcinogenic activities of two SRC-II distillates were compared. Distillates boiling from 800°F to 850°F and more than 859°F each contained about 1800 ppm B(a)P. In addition, analysis showed that the 800° to 850° distillate contained numerous other known carcinogens. Yet the greater than 850° distillate was 3 to 5 times more carcinogenic than the 800°to 850° one. Chronic skin painting assay results also demonstrate a higher activity for the greater than 850° distillate (DOE 1987).

Assay of PAH fractions from these two distillates resulted in similar differences in carcinogenic activity, even though they contained the same quantities of B(a)P. This provides additional evidence that B(a)P activity is inhibited in complex mixtures. Further evidence was obtained when the PAH fraction for the greater than 850° distillate was further fractionated using high-performance liquid chromatography. This technique resulted in a fraction greatly enriched in B(a)P and greatly reduced in the number of PAH. Assay of this B(a)P-enriched fraction showed that its carcinogenic activity was greater than that of the parent PAH fraction and approached that of pure B(a)P (DOE 1987).

Because the results of the various studies indicated that the carcinogenic activity of many complex mixtures was not correlated with B(a)P levels and that the activity of B(a)P was inhibited when tested as a part of a complex mixture, Mahlum et al. 1984, as cited in DOE (1987), directly tested the influence of coal liquids on the skin tumor-initiating activity of B(a)P. In the first experiment, the carcinogenic activity of B(a)P in acetone was compared with that of B(a)P administered in a broad boiling range (300 to 850°F) liquid. Initiating doses of either 10 or 50 μg of B(a)P were used. With both doses, the activity of B(a)P was significantly lower when applied in the coal liquid. These results clearly demonstrated the inhibiting effect of the mixture on B(a)P tumorigenic activity (DOE 1987).

Scientists at PNL recently completed an experiment to determine if the inhibition of B(a)P activity was associated with materials having specific boiling ranges. Mahlum et al. 1984, as cited in DOE (1987), tested the effect of discrete boiling point cuts prepared from the 300 to 850°F liquid noted above, on the expression of B(a)P activity by applying a 25-μg dose of B(a)P to mouse skin in acetone:methylene chloride vehicle or in acetone:methylene chloride containing 5 μg of the distillate of interest. The results of this experiment indicated that a distillate boiling from 300 to 700°F had little effect on B(a)P expression. However, all distillates with boiling ranges of 700, 750 to 800, 800 to 850, and more than 850°F markedly reduced the carcinogenic activity of B(a)P, probably as a result of altered metabolism (DOE 1987).

In other PNL experiments, the influence of complex mixtures on metabolism and DNA binding of B(a)P was examined. The results of these experiments are consistent with the skin tumor experiments in that both the rate of metabolism and the extent of binding of B(a)P to DNA are inhibited by the high boiling coal distillates. These results demonstrate unequivocally that the

B(a)P content of mixtures is not well correlated with carcinogenic activity (or microbial mutagenic activity). Further, PNL data suggest that the failure of B(a)P levels to correlate with carcinogenic activity is caused by an inhibition of expression of B(a)P activity by other components of the mixture (DOE 1987).

2.3.2   Carbon Dioxide

Carbon dioxide is a colorless, odorless gas. Its molecular formula is CO_2 and its molecular weight is 44.01. It has a Hazard Rating of 1 (Sax and Lewis 1989). The TLV, on a TWA basis, is 5,000 ppm. The TLV, on a STEL basis, is 30,000 ppm (ACGIH 1988). Common synonyms for carbon dioxide are carbonic acid gas and carbonic anhydride.

2.3.2.1   Summary

Carbon dioxide is one of the two major products of combustion (i.e., CO_2 and H_2O vapor). Human metabolic activity accounts for most of the indoor CO_2 concentrations.

It is recognized that repeated daily exposure at 0.5 to 1.5% inspired carbon dioxide at 1 atmosphere pressure is well tolerated by normal individuals (HSDB 1990).

Two percent carbon dioxide in inhaled air increases pulmonary ventilation 50%. Dizziness, headache, confusion and dyspnea occur at 5% concentration carbon dioxide. Eight to ten percent causes severe headache, sweating, dimness of vision and tremor; consciousness is lost after 5 to 10 minutes (HSDB 1990).

2.3.2.2   Health Effects

Carbon dioxide is one of the two major products of combustion (i.e., CO_2 and H_2O vapor). Human metabolic activity accounts for most of the indoor CO_2 concentrations.

General Health Effects - It is recognized that repeated daily exposure at 0.5 to 1.5% inspired carbon dioxide at 1 atmosphere pressure is well tolerated by normal individuals (HSDB 1990).

Two percent carbon dioxide in inhaled air increases pulmonary ventilation by 50%. Dizziness, headache, confusion and dyspnea occur at 5% concentration of carbon dioxide. Eight to ten percent causes severe headache, sweating, dimness of vision and tremor; consciousness is lost after 5 to 10 minutes (HSDB 1990).

In a normal person, inhalation of 1.6% carbon dioxide air approximately doubles the respiratory minute volume and at 5% almost triples it. A concentration of 10% produces unbearable dyspnea after a few minutes, and continued exposure results in vomiting, disorientation and hypertension (HSDB 1990).

Adding 1% carbon dioxide to air increased the human pulmonary ventilation rate by 37 and 7% on the ground and under a pressure simulating a 5000-m altitude, respectively. Blood flow to the brain increased at 3 but not 1% carbon dioxide. Carbon dioxide at 0.5% or 1% stimulated hyperventilation to a degree which prevented a decrease in the psychomotor performance at a simulated 5800-m, but not at 5000-m, altitude (HSDB 1990).

Long-term exposure to levels between 0.5 and 1%, as may occur in submarines, is likely to involve increased calcium deposition in body tissues, including the kidney (HSDB 1990).

2.3.3 Carbon Monoxide

Carbon monoxide (CO) is a colorless and odorless gas. Its molecular formula is CO and its molecular weight is 28.01. It has a Hazard Rating of 3 (Sax and Lewis 1989). Common synonyms for carbon monoxide are carbonic oxide, and carbon oxide.

2.3.3.1 Summary

Carbon monoxide causes hypoxia (oxygen starvation) and, at sufficiently high concentrations, asphyxiation. By binding to hemoglobin (Hb) in the red blood cells and forming COHb, CO displaces oxygen at the binding sites of the Hb molecule that normally serves as the main carrier for transporting oxygen to the cells in the body. The ability of the blood to carry oxygen is thus reduced. Relatively low atmospheric concentrations of CO can result in significant oxygen deprivation, as reflected by high COHb levels, because CO has an approximately 220-fold higher affinity to Hb than oxygen. A CO concentration of 550 ppm, for example, can lead to COHb levels of 20% within 1 hour.

The central nervous system, the cardiovascular system, and the liver are tissues most sensitive to CO-induced hypoxia. Manifestations of central nervous system hypoxia become apparent at COHb concentrations of 2 to 5%. They include loss of alertness; impaired perception and judgement; loss of coordination; reduced performance, vigilance, and concentration; drowsiness; confusion; and, at sufficiently high concentrations and sufficiently long exposures, coma and death (DOE 1987).

Cardiovascular hypoxia manifests itself at COHb levels as low as 3%. It causes a decrease in exercise time to produce angina pectoris, increased incidence of myocardosis (degeneration of heart muscle), and increased probability of heart failure in susceptible patients (DOE 1987).

2.3.3.2   Health Effects

General Health Effects - Carbon monoxide causes hypoxia (oxygen starvation) and, at sufficiently high concentrations, asphyxiation. By binding to hemoglobin (Hb) in the red blood cells and forming COHb, CO displaces oxygen at the binding sites of the Hb molecule that normally serves as the main carrier for transporting oxygen to the cells in the body. The ability of the blood to carry oxygen is thus reduced. Relatively low atmospheric concentrations of CO can result in significant oxygen deprivation, as reflected by high COHb levels, because CO has an approximately 220-fold higher affinity to Hb than oxygen. A CO concentration of 550 ppm, for example, can lead to COHb levels of 20% within 1 hour. By comparison, normal physiological values for nonsmokers range from 0.3 to 0.7% (Coburn et al. 1963; as cited in DOE 1987). Cardiovascular patients can experience aggravation of their symptoms at COHb levels of 2.5 to 4%. A 1-hour exposure to 1,500 ppm is potentially lethal for healthy persons. Chronic exposure to lower concentrations, as experienced by smokers and traffic-regulating policemen, stimulates the body to compensate somewhat by producing more red blood cells (erythrocytes) and more Hb. A threshold might exist at a COHb concentration at which adaptation can no longer compensate (NRC 1977, as cited in DOE 1987).

The central nervous system, the cardiovascular system, and the liver are tissues most sensitive to CO-induced hypoxia. Manifestations of central nervous system hypoxia become apparent at COHb concentrations of 2 to 5%. They include loss of alertness; impaired perception and judgement; loss of coordination; reduced performance, vigilance, and concentration; drowsiness; confusion; and, at sufficiently high concentrations and sufficiently long exposures, coma and death (DOE 1987).

Cardiovascular hypoxia manifests itself at COHb levels as low as 3%. It causes a decrease in exercise time to produce angina pectoris, increased incidence of myocardosis (degeneration of heart muscle), and increased probability of heart failure in susceptible patients (DOE 1987).

Populations at special risk include fetuses; patients with certain health problems, such as cardiovascular or chronic respiratory (bronchitis, emphysema, or asthma) diseases; individuals under the influence of drugs or alcohol; the elderly; and persons not adapted to high altitude who are exposed to CO at high altitudes. A combination of any two or more risk factors increases the risk accordingly. Table 2.11 shows threshold COHb concentrations (DOE 1987).

It is expected that individuals with predisposing illnesses or physiological conditions that limit pulmonary gas exchange or oxygen transport to body tissues would be more susceptible to the hypoxia effects of CO. People with angina pectoris, peripheral vascular disease, and other cardiovascular diseases appear to be at greatest risk from low-level exposures to CO (EPA 1984c, as cited in DOE 1987). Anderson et al. (1973, as cited in DOE (1987), observed that angina pectoris patients can be affected by COHb levels from 2.9 to 4.5%. Aronow et al. (1974), as cited in DOE (1987), suggest that patients with peripheral vascular disease can be at risk from low-level CO exposures.

Infants born to women who have survived acute exposure to high concentrations of carbon monoxide while pregnant often display neurological sequelae and there may be gross damage to the brain. Persistent low levels of carboxyhemoglobin in fetuses of women who smoke may also reduce the infants' mental abilities (HSDB 1990).

TABLE 2.11.   Threshold COHb Concentrations

Category	Threshold COHb %	Effects
A	0.3 to 0.7	Physiological norm for nonsmokers
B	2.5 to 3.0	Cardiac function decrements in impaired individuals; blood flow alterations; and, after extended exposure, changes in red blood cell concentration
C	4.0 to 6.0	Visual impairments, vigilance decrements, reduced maximal work capacity. Norm for smokers
D	10.0 to 20.0	Slight headache, lassitude, breathlessness from exertion, dilation of blood cells in the skin, abnormal vision, potential damage to fetuses
E	20.0 to 30.0	Severe headaches, nausea, abnormal manual dexterity
F	30.0 to 40.0	Weak muscles, nausea, vomiting, dimness of vision, severe headaches, irritability, and impaired judgement
G	50.0 to 60.0	Fainting, convulsions, coma
H	60.0 to 70.0	Coma, depressed cardiac activity and respiration, sometimes fatal

2.3.4 <u>Environmental Tobacco Smoke</u>

Tobacco smoke, with about 38000 constituents (EPRI 1985, as cited in DOE 1987), can contribute substantially to indoor air pollution under conditions of heavy smoking and poor ventilation. Constituents of the gas phase of tobacco smoke include CO, CO_2, oxides of nitrogen, ammonia, volatile N-nitrosamines, hydrogen cyanide and cyanogen, volatile hydrocarbons, volatile alcohols, and volatile aldehydes and ketones. The particulate phase comprises nicotine, water and "tar." The term "tar" refers to nonvolatile N-nitrosamines, aromatic amines, alkanes, alkenes, tobacco isoprenoids, benzenes, and naphthalenes, poly aromatic hydrocarbons, N-heterocyclic hydrocarbons (aza-arenes), phenols, carboxylic acids, metallic constituents, radioactive substances, agricultural chemicals, and tobacco additives. Many of these chemicals have been found to be toxic or carcinogenic in humans and/or laboratory animals (DOE 1987). Environmental tobacco smoke has a Hazard Rating of "D" (Sax and Lewis 1989).

2.3.4.1 <u>Summary</u>

The harmful effects of smoking on the smoker are well known and documented. However, health effects of chronic involuntary exposure of nonsmokers to environmental tobacco smoke (ETS) at concentrations that generally are well below recommended TLVs for the individual chemicals, are still controversial. Acute effects on susceptible individuals, mainly temporary irritation of the eyes and mucous membranes of the upper respiratory tract, and headaches, are generally recognized. In contrast, results of epidemiological studies linking involuntary exposure of ETS to an increased incidence of 1) cancer in nonsmoking spouses, 2) morbidity (especially respiratory diseases) in infants and children of smoking parents (especially smoking mothers), and 3) aggravation of symptoms in patients with cardiovascular and chronic obstructive lung diseases (COLD) are still heatedly debated. Even within the biomedical community, opinions vary widely.

In view of the difficulties of 1) making meaningful concentration measurements, 2) developing standards for normalizing exposures occurring under widely varying conditions, 3) conducting "perfect" epidemiological studies, 4) developing a consensus regarding the interpretation of results of such studies, and 5) resolving the existing disagreements and controversy in the biomedical community regarding these issues, no meaningful risk assessments for ETS exposures and quantifications of health effects can be made at this time (DOE 1987).

Several of the studies that have been conducted to address this controversial problem are discussed in Section 2.3.4.2, "Health Effects". Against the evidence from these and some other epidemiological studies stands criticism of the methods, statistics, and data interpretation. Also there are several other studies with ambiguous or negative studies. Several of these other studies are listed in DOE (1987).

Over the past several months, the EPA has been reviewing the effects of ETS. In the past, the EPA has used previous estimates by private researchers on the damage caused by ETS. Such estimates have ranged from 12 to 5,200 lung cancer deaths annually in the United States and as many as 46,000 deaths overall, if illnesses such as heart disease and respiratory ailments are included. The findings of the EPA's review are in draft form, which have yet to be reviewed by the EPA's Scientific Advisory Board.

2.3.4.2   Health Effects

General Health Effects - The harmful effects of smoking on the smoker are well known and documented. However, health effects of chronic involuntary exposure of nonsmokers to ETS at concentrations that generally are well below recommended TLVs for the individual chemicals, are still controversial. Acute effects on susceptible individuals, mainly temporary irritation of the eyes and mucous membranes of the upper respiratory tract, and headaches, are generally recognized. In contrast, results of epidemiological studies linking involuntary exposure of ETS to an increased incidence of 1) cancer in nonsmoking spouses, 2) morbidity (especially respiratory diseases) in infants and children of smoking parents (especially smoking mothers), and 3) aggravation of symptoms in patients with cardiovascular diseases and COLD, are still heatedly debated. Even within the biomedical community, opinions vary widely. Based on statements from experts at recent workshops, the position of the majority of investigators in this field appears to be that the available evidence for significant health effects of involuntary chronic exposure to ETS is ambiguous and incomplete. ETS health effects are considered minor and negligible, but this opinion is not universally shared. Almost all investigators do agree on the need for further studies, particularly in defining passive smoke exposure conditions (DOE 1987).

Quantitative Risk Estimates - In view of the difficulties of 1) making meaningful concentration measurements B(a)P standards for normalizing exposures occurring under widely varying conditions, 3) conducting "perfect" epidemiological studies, 4) developing a consensus regarding the interpretation of results of such studies, and 5) resolving the existing disagreements and controversy in the biomedical community regarding these issues, no meaningful risk assessments for ETS exposures and quantifications of health effects can be made at this time (DOE 1987).

Acute Health Effects - Many deleterious effects of tobacco smoke on the health of smokers have long been established beyond reasonable doubt and are generally recognized by the scientific and medical communities. However, considerable differences exist between active and passive exposure to tobacco smoke (exposure conditions and duration, quality and quantity of inhaled combustion products). Therefore, unqualified extrapolation from the health effects of active smoking to those of passive smoking would not be a valid scientific approach. The extrapolation would also be very tenuous when it is realized that those most heavily exposed to ETS are the smokers themselves (DOE 1987).

Measurable quantities of physiologically active smoke components (e.g., CO, nicotine and its metabolite cotinine) are absorbed by involuntary smokers. Many nonsmokers are bothered by tobacco smoke (National Clearinghouse for Smoking and Health 1975, as cited in DOE 1987) and/or experience irritation of the eyes and of the nasopharyngeal mucosa, headache, and cough (Speer 1968; Barad 1979; Weber et al. 1976 1979; Hugod et al. 1978; Johansson 1976; Johansson and Ronge 1965, all as cited in DOE 1987). Age and sensitivity to tobacco smoke components may affect the acute reaction(s) of nonsmokers in addition to exposure conditions.

Several investigators reported acute physiological effects in the form of increased heart rate and blood pressure in healthy subjects exposed to cigarette smoke (Luquette et al. 1970; Rummel et al. 1975, as cited in DOE 1987), and small decreased in maximal aerobic capacity, exercise time to exhaustion, and maximal oxygen consumption (Aronow and Cassidy 1975; Gliner et al. 1975; Raven et al. 1974, all as cited in DOE 1987). However, these effects were minor and transitory and, therefore, probably inconsequential. There were no effects on ventilatory lung functions such as lung volume, maximal expiratory flow volume, single-breath nitrogen washout, submaximal exercise (Pimm et al. 1978; Gliner et al. 1975, as cited in DOE/BPA 1978) and on lung functions during intermittent bicycle ergometer performance (Shephard et al. 1979, as cited in DOE 1987).

Chronic Health Effects - Of greater concern than the nuisance and/or transient acute effects are the potentially more serious chronic health effects of passive smoke exposure. Observations in recent years have focused concern mainly on three areas (DOE 1987):

- effects of chronic exposure to ETS on respiratory functions in infants and children

- effects of chronic exposure to ETS on patients with predisposing diseases

- effects of chronic exposure to ETS on the development of malignant tumors

Effects on Respiratory Functions of Infants and Children - Colley (1974), as cited in DOE (1987), observed a higher incidence of respiratory illness in children with smoking parents. However, this might have been an indirect effect of ETS because of an even stronger relationship between cough and mucus production in the smoking parents, resulting in greater disease-spreading capacity of these parents, and respiratory infections in their children. Bland et al. (1978), as cited in DOE (1987), described similar findings. In another study, Colley et al. (1974) and Leader et al. (1976), as cited in DOE (1987), investigated the incidence of bronchitis and pneumonia in 2,205 children during the first 5 years of life. The authors found a cause-effect relationship between these incidences and parental smoking habits

(number of parents smoking; number of cigarettes smoked), but only for the
first year of life. However, in a different paper, Cederlof and Colley
(1974), as cited in DOE (1987), pointed out that, "when parents' respiratory
symptoms were taken into account, exposure of the child to cigarette smoke
generated by the parents' smoking had little if any effect upon the child's
respiratory symptoms." Lebowitz and Burrows (1976) and Schilling et al.
(1977), as cited in DOE (1987), also came to the conclusion that the effects
of parental smoking on children were insignificant when parental symptoms were
taken into account.

On the other hand, Rentakallio (1978a,b), as cited in DOE (1987), in a
study with 12,000 Finnish children, found a significantly (P<0.001) higher
morbidity, and increased (P<0.001) and longer hospitalization, mostly from
respiratory diseases, in children of smoking mothers. This increased
morbidity manifested itself during the first 5 years of life and was most
pronounced during the first year of life. Harlap and Davies (1974), as cited
in DOE (1987), reported a dose-response relationship between maternal smoking
and hospital admissions of infants for bronchitis and pneumonia, but only
between the sixth and ninth months of life. Tager et al. (1976, 1979, 1983),
as cited in DOE (1987), described a decline in expiratory flow rates in
children as a function of the number of parents smoking and the number of
cigarettes smoked. Weiss et al. (1980), as cited in DOE (1987), observed a
significant linear relationship between parental smoking and persistent
wheezing and decreased mean forced midexpiratory flow. Ware et al. (1984), as
cited in DOE (1987), found that maternal cigarette smoking was associated with
increases of 20 to 35% in the incidences of eight respiratory illnesses and
symptoms investigated at two successive annual examinations of 10,106 white
children living in six cities. Paternal smoking was associated with smaller
but still substantial increases. Illness and symptom incidences were linearly
related to the number of cigarettes smoked by the child's mother, and were
higher for children of current smokers than for children of ex-smokers. The
association between maternal smoking status and childhood respiratory
illnesses and symptoms wee reduced, but not eliminated, by adjustment for
parental illness history. Levels of forced expiratory volume in 1 second
(FEV_1) were significantly lower for children of current smokers that for

children of nonsmokers at both examinations, and highest for children of
ex-smokers. Levels of forced vital capacity (FVC) were lower for children of
nonsmokers than for children of current smokers at both examinations, but the
difference was statistically significant only at the first examination. Both
the increase in mean FVC and the decrease in mean FEV_1 among children of
current smokers were linearly related to daily cigarette consumption. None of
the respiratory illnesses and symptoms studies were significantly associated
with exposure to gas cooking in the child's home. The results suggest a
causal effect of sidestream cigarette smoke on increased respiratory illness
and reduced FEV_1 values in preadolescent children (DOE 1987).

Gardner et al. (1984), as cited in DOE (1987), monitored 131 infants
from birth through the first year of life to investigate effects of social and
familial factors on respiratory diseases. The authors found a significant
relationship between pneumonia and parental smoking, especially smoking by
mothers at home. Analysis of questionnaire data and lung function tests by
Tashkin et al. (1984), as cited in DOE (1987), on children revealed that
passive exposure to maternal smoking affected airways of boys aged 7 to 11. A
total of 971 white, non-hispanic, nonsmoking, nonasthmatic children were
divided into three categories related to parental smoking status: 1) at least
mother smokes, 2) only father smokes, and 3) neither parent smokes. Prediction equations for several indices of forced expired volume and flow were
derived separately for boys and girls 7 to 17 years of age. Analysis of
variance was performed separately on younger (aged 7 to 11) and older (12 to
17) children of each sex. Among younger male children, residual values were
significantly lower in the maternal smoking category than in the other two
household categories for maximal flow after exhalation of 25% of FVC; no
differences were noted between the paternal smoking only and nonsmoking
household categories. The differences between maternal and paternal smoking
effects are probably explained by the longer daily exposure of children of
nonemployed smoking mothers. A trend toward similar results was found in
older male children. Among females, forced expiratory flow during the middle
half of the FVC and maximal flow after exhalation of 75% of FVC were
significantly lower in relation to maternal smoking in older children only.
Analysis of Variance (ANOVA) revealed no decrement in lung function in

relation to passive smoking among children with asthma or bronchitis (n=138). Chi-square analysis showed no differences in the frequency of respiratory symptoms among children in the different passive exposure categories. The authors noted that the apparent effect on lung function of heavy exposed older girls is more likely to be confounded by selective under-reporting of active smoking (DOE 1987).

A review by Weiss et al. (1983); as cited in DOE 1987, reflects the current state of knowledge on the effects of chronic exposure to ETS in infants and children and in patients with predisposing diseases. The authors point to the relatively large body of data relating parental (particularly maternal) cigarette smoking to the occurrence of both acute respiratory illnesses and chronic respiratory symptoms in children. The effect appears to be greatest early in life and cannot be separated from in utero exposure. While data liking parental smoking to lower levels of pulmonary function are all cross-sectional and less conclusive, the magnitude of the direct effect of passive smoke exposure is likely to be relatively small (from 1 to 5% reduction in maximally obtained lung-function level in exposed children). The important effects of passive smoke exposure in childhood are twofold: the slight reduction n pulmonary function may predispose an individual to increased risks from environmental agents later in life; and having a parent who smokes substantially increased the likelihood that a child will become a smoker. Involuntary smokers are exposed to a quantitatively smaller and qualitatively different (but not necessarily less hazardous) smoke exposure than active smokers. Quantitation of exposure is particularly difficult in physiological and epidemiological studies. Acute physiological studies have documented minimal physiological changes in healthy subjects. However, individuals with heart or lung disease may be affected more. Data on adults are insufficient to allow a quantitative estimate of effects. The authors state, as many others have done, that further research is needed to confirm findings on passive smoking (DOE 1987).

Reviewing the link between parental smoking habits and respiratory symptoms in offspring, Holt and Turner (1984), as cited in DOE (1987), conclude that infants whose parents smoke experience a greater frequency of respiratory infections, particularly in the first year of life. Increased wheezing, exacerbation of asthma, and reduced lung function in older children have also been reported. These effects appear to be more strongly related to maternal than to paternal smoking, especially when infants are involved (DOE 1987).

According to Bake (1984), as cited in DOE (1987), data strongly suggest an association between exposure to ETS in the home and a small reduction in ventilatory lung function in adults exposed more than 15 to 20 years. However, the potential magnitude of ETS exposure is uncertain because no study has considered the combined effects of various related elements. Bake sees indications that, all other factors being equal, exposure to ETS would probably also affect lung function in children. However, a review of various studies show uncertainty regarding parental smoking effects and only minimal consequences in children (DOE 1987).

Effects on Patients with Predisposing Diseases - There is evidence that ETS-induced COHb levels can reduce the exercise duration required to provoke angina pectoris in patients with coronary artery disease (Aronow and Isbell 1973; Aronow et al. 1974a,b; Anderson et al. 1973, all as cited in DOE 1987).

Patients with COLD experienced a reduction in mean exercise time until onset of marked dyspnea from 219 to 147 seconds. However, expected effects on blood pressure, heart rate, arterial pO_2/pCO_2 or pH were not detected, thus leaving unresolved the mechanism of early dyspnea induction (Aronow et al. 1977, as cited in DOE 1987).

It is also evident that allergic individuals can be more sensitive to many environmental pollutants/irritants, including tobacco smoke, but it remains unresolved whether this constitutes a true allergy following a specific sensitization to cigarette smoke (Taylor 1974, as cited in DOE 1987). Parental smoking is also a significant exacerbating factor in childhood asthma (O'Connel and Logan 1974, as cited in DOE 1987).

Effect on Development of Malignant Tumors - Of greatest concern is the question of whether or not chronic exposure to ETS presents a (lung) cancer

risk to exposed nonsmokers. The most frequently cited and debated epidemiological studies are those of Hirayama (1981) and of Trichopoulos et al.
(1981), as cited in DOE (1987).

From 1966 to 1979, Hirayama (1981), as cited in DOE (1987), investigated
mortality records in 29 health center districts, studied 91,540 nonsmoking
wives 40 years of age or older, and assessed standardized mortality rates for
lung cancer according to the smoking habits of their husbands. He found that
wives of former smokers and of smokers of fewer than 20 cigarettes per day had
a relative risk factor, or risk ratio (RR), of 1.6. Wives of heavy (more than
20 cigarettes per day) smokers had a relative risk of 2.1. No significant
effects on the incidence of other forms of cancer or other diseases were
apparent at that time, but there was a statistically insignificant tendency
toward a higher risk of developing emphysema and asthma in the nonsmoking
wives of heavy smokers. Updating his findings 2 years later, Hirayama (1983),
as cited in DOE (1987), noted that of 429 women who died from lung cancer
during 16 years of follow-up (1966 to 1981), 303 had been nonsmokers; 200 of
these deaths occurred among the 91540 nonsmoking married women whose husband's
smoking habits were known. Standardized mortality ratios of lung cancer in
nonsmoking women were 1.00, 1.36, 1.42, 1.58, and 1.91 when husbands were
nonsmokers, ex-smokers, and daily smokers of more than less than 15, 15 - 19,
and more than 19 cigarettes per day, respectively. Based on the extent of the
husbands' smoking, trends of increased risk among nonsmoking women were
similar for different age groups, occupational groups, and duration of observation. Smoking by the husband was the only factor found to increase the risk
of lung cancer in nonsmoking wives. A significantly increased risk of cancer
of the paranasal sinuses in nonsmoking wives, related to the amount of
husbands' smoking, was also observed (DOE 1987).

Trichopoulos et al. (1981), as cited in DOE (1987), questioned 51 women
on lung cancer and 163 other hospital patients in Greece on their and their
husbands' smoking habits. Forty of the lung cancer patients and 149 of the
other patients were nonsmokers. Among the nonsmoking women, there was a
statistically significant difference between the cancer patients and the other
patients with respect to their husband's smoking habits. The relative risk of

lung cancer associated with a smoking husband was 2.4 for smokers of less than 20 cigarettes per day and 3.4 for heavier smokers. The authors updated their findings 2 years later (Trichopoulos et al. 1983, as cited in DOE 1987). While they recognized that their study (1981) had been criticized for the small number of subjects, lack of histological confirmation, and hospital differences, the authors point out that there were twice the number of cases and 50% more controls at the conclusion of the study, with the results remaining substantially the same. For a total of 77 cases and 225 nonsmoking female controls, the RR of lung cancer for never-married women or those with husbands who were nonsmokers or ex-smokers for 5 to 20 years (24 cases, 109 controls) was 1.0; for spouses of men who were ex-smokers for less than 5 years (15 cases, 35 controls), the RR was 1.9; for spouses of men currently smoking for 1 to 20 cigarettes per day (24 cases, 56 controls), the RR was 2.4; and for women whose husbands currently smoked 21 or more cigarettes per day (14 cases, 25 controls), the RR was 3.4. The authors comment that the relatively low RR and the many potential sources of bias preclude any single study from providing conclusive evidence, but that the similarity of results from different studies in different populations will permit valid conclusions regarding effects of ETS on lung cancer incidence in nonsmokers (DOE 1987).

Correa et al. (1983), as cited in DOE (1987), investigated the smoking habits of parents and spouses in a case-control study involving 1,338 lung cancer patients and 1,393 control subjects in Louisiana. Nonsmokers married to heavy smokers had an increased risk of lung cancer, as did subjects who had mothers who smoked. There was no association between lung cancer risk and paternal smoking. The association with maternal smoking was found only in smokers and persisted after controlling for variables, which indicated active smoking. The authors state that it is not clear whether the results reflect a biological effect associated with maternal smoking or the inability to control adequately for confounding factors related to active smoking (DOE 1987).

In a German Study (Knoth et al. 1983, as cited in DOE 1987) of 792 patients of both sexes with brochogenic carcinoma, there were 59 female patients, 39 of them nonsmokers. Of the nonsmoking female patients, 61.5% had lived with smokers, triple the number anticipated based on the smoking habits of men in their respective age groups. Neither occupational exposure nor hereditary factors accounted for the excess. Hence, exposure to ETS may be the most obvious interpretation of the high percentage of nonsmokers among the female patients with bronchogenic carcinoma. In addition, the percentage of squamous cell and small-cell carcinoma, the typical smokers' carcinomas, among the nonsmoking wives (66.6%) was not significantly lower than that among the females who smoked (80%) (DOE 1987).

Gillis et al. (1984), as cited in DOE (1987), analyzed ETS exposure, smoking and mortality rates for 8,128 subjects who attended a multiphasic screening unit. The subjects were between 45 and 64 years of age, and 97% were male/female partnerships. The prevalence of infected mucus, persistent mucus, dyspnea, and hypersecretion was slightly higher in ETS-exposed non-smokers than in controls (nonexposed nonsmokers). Male lung cancer mortality was 4/10,000 in controls, 13/10,000 for ETS-exposed nonsmokers, 22/10,000 in smokers, and 24/10,000 in ETS-exposed smokers. No such trend was noted among the women. Overall myocardial infarction mortality was slightly higher in the ETS-exposed group, but no other differences in smoking-related mortality were found (DOE 1987).

Preliminary results of a case-control study by Sandler et al. (1985); as cited in DOE (1987) indicate that the overall cancer risk for passive smokers rose steadily and significantly with the number of household members who smoked. These findings are based on 369 patients and 409 control subjects (DOE 1987).

Miller (1984a), as cited in DOE (1987), in an epidemiology study conducted from 1975 to 1976 in Erie County, Pennsylvania, observed a significantly increased cancer mortality in nonemployed wives chronically exposed to ETS compared to nonemployed wives with little or no ETS exposure. Several errors in Miller's first publication were subsequently corrected (Miller 1984b, as cited in DOE 1987). The corrections did not significantly change the originally reported results.

Lawrence and Paulson (1983), as cited in DOE (1987), voiced interesting thoughts on cancer risk from low concentrations of ETS. They state that one method for establishing safe dose levels for such low ETS exposure is via

extrapolation of data from high-dose experiments. A linear extrapolation
technique was applied to data from a 1954 prospective epidemiological study on
deaths from lung and bronchial cancer among 291,000 veterans. Most smokers
had begun smoking before the age of 25. Mortality was observed for 8.5 years
after questioning the participants of the study. To simulate conditions in
animal experiments, the analysis was restricted to data on respondents who
never quit smoking and smoked for most of their adult lives. The extrapolation yielded a practically safe dose of 0.005 cigarettes/day, a level of
environmental exposure that was exceeded in nonsmokers in nearly all cases.
The authors observed that there is a large discrepancy between the low levels
to which involuntary exposure to other carcinogens are increasingly regulated
and controlled, and levels of passive exposure to cigarette smoke apparently
considered acceptable. Because there is no proof that mainstream smoke is
more hazardous than sidestream smoke, the extrapolation may be questionable.
Also, linear extrapolation may not be the most conservative method (DOE 1987).

Against the evidence from these and some other epidemiological studies
stands criticism of their methods, statistics, and data interpretation. Also,
there are several other studies with ambiguous or negative studies. Several
of these other studies are listed in DOE (1987).

2.3.5 Nitrogen Oxides

Oxides of nitrogen technically include NO, NO_2, nitrous oxide (N_2O),
nitrogen trioxide (OONO), dinitrogen trioxide (N_2O_3), dinitrogen tetraoxide
(N_2O_4), and dinitrogen pentoxide (N_2O_5). All of these compounds, as well as
their secondary reaction products (e.g., nitrate aerosols) can affect human
health. However, only NO and NO_2 are of practical importance as indoor air
pollutants. Both compounds are produced from atmospheric nitrogen and oxygen
in the course of the combustion process. The biological effects of NO_2 have
been extensively studied during the past 30 years (DOE 1987).

NO_2 Noncarcinogenic Effects - The RfD is based on the assumption that
thresholds exist for certain toxic effects such as cellular necrosis, but may
not exist for other toxic effects such as carcinogenicity. In general, the
RfD is an estimate of a daily exposure to the human population (including
sensitive subgroups) that is likely to be without an appreciable risk of

deleterious effects during a lifetime. The RfDo for NO_2 is 1E+0 mg/kg/day, with methemoglobinemia as the critical effect. The NOAEL is 10 ppm of drinking water. The LOAEL is 11-20 ppm (IRIS 1990).

This is an epidemiologic study on the formation of methemoglobinemia in infants who routinely consumed milk prepared from water containing various levels of nitrate. The study analyzed all cases of infant methemoglobinemia occurring in 37 United States states irrespective of date of occurrence or type of water supply. Nitrate (nitrogen) content ranged from 10 ppm to > 100 ppm. No incidences of methemoglobinemia were found to occur in drinking water containing less than 10 ppm nitrate (nitrogen). Therefore, a NOAEL of 10 ppm was derived. Several more recent epidemiologic studies support Walton's (1951), as cited in IRIS (1990), threshold for infant methemoglobinemia (NAS 1977; Winton 1971; Calabrese 1978, as cited in IRIS (1990).

The confidence in the RfDo is high. Confidence in the study is high because the NOEL is determined in the known sensitive human population. The database contains several recent supporting epidemiologic studies for the critical effect in the sensitive population (infants); therefore, a high confidence rating is given to the database. High confidence in the RfDo follows (IRIS 1990).

Currently there is no RfDi for NO_2. A risk assessment for NO_2 is under review by an EPA work group.

General Health Effects - There are basically three mechanisms of NO_x toxicity (DOE 1985, as cited in DOE 1987):

- Formation of highly reactive free radicals by oxidation of unsaturated fatty acids such as lecithin, a major component of cell membranes. Free radicals can 1) interfere with the chemistry and physiology of these membranes; 2) change structural proteins (e.g., elastin and collagen), thereby affecting the structural and mechanical integrity of lung tissue; and 3) react with Hb to form methemoglobin, thereby reducing the O_2-carrying capacity of the blood.

- Formation of highly ionized acid in the respiratory tract, which probably accounts for the acute irritation.

- Formation of potentially carcinogenic nitrosamines in the respiratory tract and especially in the acid milieu of the stomach, which might also contribute to liver dysfunction.

Nitric oxide is relatively nontoxic at normally encountered concentrations, but persists longer indoors than outside and has a longer indoor half-life than NO_2. Therefore, it cannot be ignored as an indoor air pollutant. Nitrogen dioxide affects the respiratory system not unlike ozone (O_3), albeit to a lesser degree (DOE 1987).

Both NO and NO_2 bind to Hb to form methemoglobin at similar rates, thus reducing the O_2-carrying capacity of the blood. It can be assumed that NO at 3 ppm is physiologically similar to 10 to 15 ppm of CO (Case et al. 1979; as cited in DOE 1987). Nitric oxice and NO_2 can change heme by causing polycythemia. They can also cause leukocytosis and other hematological changes, and vascular membrane lesions that can result in edema (NRC 1977, as cited in DOE 1987). Nitrogen dioxide decreases the activity of acetylcholinesterase in erythrocyte membranes, increases the activity of glucose-6-phosphate dehydrogenase, increases peroxidized erythrocyte lipids, and causes significant decreases in Hb and hematocrit values (Posin et al. 1978, as cited in DOE 1987).

Oxides of nitrogen can cause acute and chronic changes in the small airways and lungs. A 4-hour exposure of rats to 2 ppm induced nonciliated cells of the small airways to differentiate into ciliated cells and mature Clara cells (Evans and Freeman 1980, as cited in DOE 1987). It also caused a proliferation of alveolar Type II cells while destroying epithelial Type I cells. Gardner et al. (1979); as cited in DOE 1987, observed significantly increased mortality in animals challenged with bacterial aerosols following NO_2 exposure to 1.5 ppm for 2 hours or to 0.5 ppm for 2 weeks.

In healthy humans, respiratory functions generally are not affected at levels of 1.5 ppm NO_2 or below. However, sensitive individuals can experience respiratory tract irritation at 0.5 ppm NO_2. An overview of effects of short-term NO_2 exposures (<3 hours) in healthy and sensitive individuals is provided in Figure 2.1.

Children and people with asthma, chronic bronchitis, and emphysema appear to be the most sensitive population groups (Table 2.12). People with

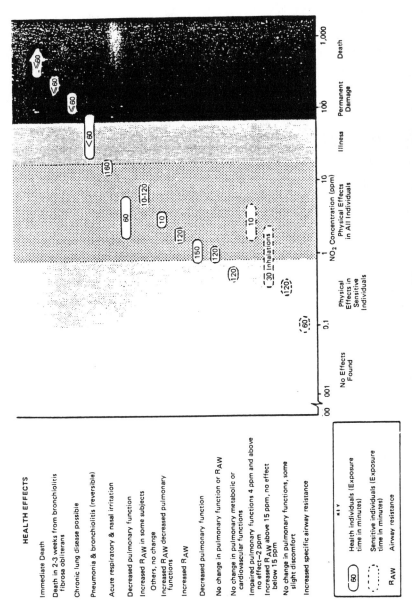

FIGURE 2.1.   Effects of Short-Term Exposures to Nitrogen Dioxide in Healthy and Sensitive Humans
SOURCE: U.S. DOE (1985), pp. 2-5, as cited in DOE (01987)

TABLE 2.12. Population Groups Sensitive to NO_2

Sensitive Group:	Children
Supporting Evidence:	Children under age 2 exhibit increased prevalence of respiratory infection when living in homes with gas stoves. Children up to age 11 exhibited increased prevalence of respiratory infections when living in gas-stove homes.
Supporting References:	Speizer et al. 1980; Melia et al. 1979
Population Estimates:	Age 0-5, 17.2 million[a]; age 5-13, 36.6 million[a]

Sensitive Group:	Asthmatics
Supporting Evidence:	Asthmatics reacted to lower levels of NO_2 than normal subjects in controlled human exposure studies.
Supporting References:	Kerr et al. 1979; Orehek et al. 1976
Population Estimates:	6.0 million[a]

Sensitive Group:	Chronic bronchitics
Supporting Evidence:	Chronic bronchitics reacted to low levels of NO_2 in controlled human exposure studies.
Supporting References:	Kerr et al. 1979; Von Nieding et al. 1971, 1973
Population Estimates:	6.5 million[a]

Sensitive Group:	Emphysematics
Supporting Evidence:	Emphysematics have significantly impaired respiratory systems. Because studies have shown that NO_2 impairs respiration by increasing airway resistance, it is reasonable to assume that emphysematics may be sensitive to NO_2.
Supporting References:	Von Nieding et al. 1971; Beil and Ulmer 1976; Orehek et al. 1976
Population Estimates:	1.3 million[a]

TABLE 2.12. (contd)

Sensitive Group: Persons with tuberculosis, pneumonia, pleurisy, hay
                          fever or other allergies.

Supporting Evidence: Studies have shown that NO_2 increases airway resis-
                          tance. Persons who have or have had these conditions
                          may be sufficiently impaired to be sensitive to low
                          levels of NO_2.

Supporting References: Von Nieding et al. 1971; Beil and Ulmer 1976; Orehek
                          et al. 1976.

Population Estimates: Unknown

Sensitive Group: Persons with liver, blood or hormonal disorders.

Supporting Evidence: NO_2 induces changes in liver drug metabolism, lung
                          hormone metabolism, and blood biochemistry.

Supporting References: Menzel 1980; Posin et al. 1978

Population Estimates: Unknown

(a) 1970 U.S. Bureau of Census and 1970 U.S. National Health Survey. All
     subgroups listed are not necessarily sensitive to NO_2 exposure at low
     levels.

SOURCE: U.S. Environmental Protection Agency 1982a; as cited in DOE 1987.
(Table by T. Namekata, Battelle Human Affairs Research Centers, Seattle,
Washington)

hay fever, liver, hematological, or homonal disorders can also be affected by
low levels of NO_2, but exposure data for the latter groups are too sparse for
recommending specific exposure limits (DOE 1987).

There is no convincing evidence of potentiating effects of NO_2 exposure
in the presence of other pollutants such as O_3, CO, or SO_2 (Table 2.13).
Increased sensitivity to the bronchoconstrictor acetylcholine following exposure to a mixture of NO_2, O_2, and SO_2, as reported by Von Nieding et al.
(1973), as cited in DOE (1987), is difficult to interpret because of 1) the

TABLE 2.13. Effects on Pulmonary Function in Subjects Exposed to NO_2 and Other Pollutants

Concentration (ppm): $0.05\ NO_2 + 0.11\ SO_2 + 0.025\ O_3$

Exposure Duration: 2 hours
Study Population: 11 healthy subjects

Reported Effects: Increased sensitivity to bronchoconstrictor as shown by increases in R_{aw}. No effect on A_2DO_2 or R_{aw} without bronchoconstrictor.

References: Von Nieding et al. 1973

Concentration (ppm): $0.50\ O_3$; $0.50\ O_3 + 0.29\ NO_2$; $0.50\ O_3 + 0.29\ NO_2 + 30\ CO$

Exposure Duration: 4 hours
Study Population: 4 healthy male subjects

Reported Effects: Minimal change in pulmonary function caused by O_3 alone. Effects not caused by NO_2 or CO.

References: Hackney et al. 1975

Concentration (ppm): $0.25\ O_3$; $0.25\ O_3 + 0.29\ NO_2$; $0.25\ O_3 + 0.29\ NO_2 + 30\ CO$

Exposure Duration: 2 hours
Study Population: 7 male subjects; some believed to be unusually reactive to irritants

Reported Effects: Minimal change in pulmonary function caused by O_3 alone. Effects Not increased by NO_2 or CO.

References: Hackney et al. 1975.

Concentration (ppm): $50\ CO + 5\ SO_2$; $4.8\ NO_2 + 50\ CO + 5\ SO_2$

Exposure Duration: - - -
Study Population: 3 subjects

Reported Effects: Increased dust retention from 50 to 76% after NO_2 was added to air containing SO_2 and CO.

References: Schlipkoter and Brockhaus 1963

TABLE 2.13. (contd)

Concentration (PPM):	0.5)$_3$; 0.5 O$_3$ + 0.5 NO$_2$ under the following conditions: 1) 25°C, 45% relative humidity (RH); 2) 30°C, 85% RH: 3) 35°C, 40% RH; 4) 40°C, 50% RH.
Exposure Duration:	Rest, 60 minutes; exercise, 30 minutes, rest, 30 minutes
Study Population:	8 young adults
Reported Effects:	Response found only for O$_3$; no greater than additive effect or interaction between O$_3$ and NO$_2$ was observed.
References:	Horvath and Folinsbee 1979

(Table by T. Namekata, Battelle Human Affairs Research Centers, Seattle, Washington)

uncertain health significance of altered sensitivity to bronchoconstrictors in healthy or sensitive subjects, 2) uncertainties from methodological differences between their techniques and those used by other investigators, and 3) the lack of confirmation of the findings by other investigators (DOE 1987).

Community epidemiological studies of NO$_2$ are summarized in Table 2.14. Because of methodological problems (i.e., use of Jacob-Hochheiser method) with the studies performed by Shy et al. (1970a,b) and Pearlman et al. (1971), both as cited in DOE (1987), in Chattanooga, a quantitative assessment of the health effects reported to be associated with NO$_2$ levels from these investigators is not possible. In addition, it is difficult to distinguish between health effects caused by NO$_2$ and effects caused by other pollutants identified in the ambient air (e.g., O$_3$, particulates, SO$_2$) at the time of the studies. These problems limit the usefulness of these studies for standard-setting purposes (DOE 1987).

TABLE 2.14. Effects of NO_2 on Pulmonary Function in Community
Epidemiological Studies

Concentration (ppm):	Median hourly 0.07 NO_2, 0.15 O_x; 0.35 NO_2. 0.02 O_x
Study Population:	205 office workers in L.A.; 439 office workers in San Francisco
Reported Effects:	No differences in most tests. Smokers in both cities showed greater changes in pulmonary function than nonsmokers.
References:	Linn et al. 1976.

Concentration (ppm):	High exposure area: 24-hour highs--0.055 and 0.035 NO_2; 1-hour mean--0.14 to 0.30 NO_2. Low exposure area: 1-hour mean--0.06 to 0.09 NO_2.
Study Population:	128 traffic policemen in urban Boston and 140 patrol officers in nearby suburbs
Reported Effects:	No difference in various pulmonary function tests
References:	Speizer and Ferris 1973; Burgess et al. 1973

Concentration (ppm):	High exposure group: estimates 1-hour maximum--0.25 to 0.51 NO_2. Annual mean 24-hour--0.051 NO_2. Low exposure group: estimated 1-hour maximum--0.12 to 0.23 NO_2. Annual mean 24-hour--0.01 NO_2.
Study Population:	Adult nonsmokers in L.A.
Reported Effects:	No differences found in several ventilatory measurements, including spirometry and flow-volume curves.
References:	Cohen et al. 1972

Concentration (ppm):	1-hour concentration at time of testing (1:00 p.m.) 0.02 to 0.19 NO_2
Study Population:	20 children 11 years of age

TABLE 2.14. (contd)

Reported Effects: During warmer part of year, NO_2, SO_2, and TSP significantly correlated with $\dot{V}_{max}$ at 25 and 50% FVC specific airway conductance. Significant correlation between each of four pollutants (NO_2, NO, SO_2, and TSP) and V_{max} at 25 and 50% FVC; but no clear delineation of specific pollutant concentrations at which effects occur.

References: Kagawa and Toyama 1975

Table by T. Namekata, Battelle Human Affairs Research Centers, Seattle, Washington.

Though some pulmonary-function effects related to NO_2 concentrations are shown in the investigation by Kagawa and Toyama (1975), as cited in DOE (1987), the results suggest that these effects were caused by a complex mixture of pollutants. Inadequate characterization of exposure to NO_2 prevents firm conclusions about the relationship between NO_2 exposure and resulting health effects (DOE 1987).

The findings of Shy et al. (1970a,b), Pearlman et al. (1971), and Kagawa and Toyama (1975), all as cited in DOE (1987), are not inconsistent with the hypothesis that NO_2, in a complex mixture with other pollutants in the ambient air, adversely affects respiratory function and illness in children. However, these findings neither provide clear evidence for a positive association between health effects and exposure to ambient NO_2, nor do they refute such an association. Little or no evidence of health effects at ambient NO_2 concentrations is provided by other community epidemiological studies (DOE 1987).

The community epidemiological studies cited and discussed above did not take into account exposure to, and the effects of, indoor air pollutants, such as NO_2 generated by the use of gas stoves. Results of community studies on effects associated with indoor NO_2 exposure involving gas stoves are summarized in Table 2.15. Major uncertainties in the studies by Melia et al. (1977, 1979) and Speizer et al. (1980), both as cited in DOE (1987), are the

TABLE 2.15.   Community Studies of NO_2 Involving Gas Stoves[a]

Concentration (ppm):   95th percentile of 24-hour average in activity room: 0.02 to 0.06 (gas); 0.01 to 0.05 (electric). Frequent peaks in one home of 0.4 to 0.6 (gas). Maximum peak- -1.0 (gas).

Study Population: 9,120 children, ages 6 to 10, six different cities, data also collected on history of illness before age 2.

Reported Effects[b]:   Significant association between history of serious respiratory illness before age 2 and use of gas stoves (P 0.01). Also, small but statistically significant decreases in pulmonary function (FEV_1 and FVC) in children from gas-stove homes.

References: Speizer et al. 1980

Concentration (ppm): NO_2 concentrations not measured at time of study.

Study Population: 2,554 children from homes using gas to cook compared to 3,204 children from homes using electricity, ages 6 to 11.

Reported Effects[b]:   Proportion of children with one or more respiratory symptoms or disease (bronchitis, day or night cough, morning cough, cold going to chest, wheeze asthma) increased in homes with gas- versus electric-stove homes (girls, P 0.10; boys, not significant) after controlling for confounding factors.

References: Melia et al. 1977

Concentration (ppm): NO_2 concentrations not measured in some homes studied for health effects.

Study Population: 4,827 children, ages 5 to 10

Reported Effects[b]:   Higher incidence of respiratory symptoms and disease associated with gas stoves (boys, P 0.02; girls, P 0.15) for residences in urban but not rural areas, after controlling for confounding factors.

References: Melia et al. 1979

TABLE 2.15. (contd)

Concentration (ppm):	Kitchens (weekly average): 0.005 to 0.317 (gas) and 0.006 to 0.168 (electric). Bedrooms (weekly average): 0.004 to 0.169 (gas) and 0.003 to 0.037 (electric).
Study Population:	808 children, ages 6 to 7
Reported Effects[b]:	Higher incidence of respiratory illness in gas-stove homes (P 0.10). Prevalence not related to kitchen NO_2 levels, but increased with NO_2 levels in bedrooms of children in gas-stove homes. Lung function not related to NO_2 levels in kitchen or bedroom.
References:	Florey et al. 1979; Goldstein et al. 1979 (both are companion papers to Melia et al. 1979)
Concentration (ppm):	Sample of households (24-hour average): 0.005 to 0.11 (gas), 0 to 0.6 (electric) and 0.015 to 0.05 (outdoors)
Study Population:	128 children, ages 0 to 5; 346 children, ages 6 to 10; 421 children, ages 11 to 15
Reported Effects:	No significant difference in reported respiratory illness between homes with gas and electric stoves in children from birth to 12 years.
References:	Mitchell et al. 1974. See also Keller et al. 1979.
Concentration (ppm):	Sample of households same as reported above, but no new monitoring reported.
Study Population:	174 children under 12 years of age
Reported Effects[b]:	No evidence that cooking mode is associated with the incidence of acute respiratory illness.
References:	Keller et al. 1979
Concentration (ppm):	See above for monitoring.
Study Population:	Housewives cooking with gas stoves, compared to those cooking with electric stoves. 146 households.

TABLE 2.15.   (contd)

Reported Effects[b]:	No evidence that cooking with gas is associated with an increase in respiratory disease.
References:	Keller et al. 1979

Concentration (ppm):	See above for monitoring.
Study Population:	Members of 441 households.
Reported Effects(b):	No significant difference in reported respiratory illness among adults in gas- versus electric-cooking homes.
References:	Mitchell et al. 1974. See also Keller et al. 1979.

Concentration (ppm):	Preliminary measurements, peak hourly, 0.25 to 0.50, maximum 1.0.
Study Population:	Housewives cooking with gas stoves, compared to those cooking with electric stoves.
Reported Effects[b]:	No increased respiratory illness associated with gas stove usage.
References:	U.S. Environmental Protection Agency 1974.

(a)   Exposures in gas stove homes were to NO_2 plus other gas combustion products.
(b)   Effects reported in published references are summarized here. However, EPA (1974), as cited in DOE, 1987, warns that considerable caution should be used in drawing firm conclusions from these studies.

(Table contributed by T. Namekata, Battelle Human Affairs Research Centers, Seattle, Washington)

Note: NO_2 TLV-TWA = 3 ppm or 6 mg/m^3 (ACGIH 1983)
EPA - recommended 1-hour average outdoor NO_2 standards <0.5 ppm or at range of 0.15 to 0.30 ppm, or annual standard in the range of 0.05 to 0.08 ppm (EPA 1982); as cited in DOE, 1987.

identification of agent(s) causing the reported health effects and, if NO_2, the exposure conditions (concentrations, average exposure time, and frequency) associated with the reported effects. Possible confounding factors that might be related to the increased incidence of respiratory illness and symptoms observed in children in gas-stove homes include humidity, socioeconomic status, and pollutants other than NO_2, such as CO and hydrogen cyanide, that are emitted during gas combustion. However, there is no evidence that gas stoves generate harmful concentrations of CO and hydrogen cyanide (HCN), and that these pollutants cause an increased incidence of respiratory diseases at typical indoor concentrations. The contribution, if any, of increased humidity to increased incidence of respiratory symptoms or diseases in buildings with gas stoves requires further investigation (DOE 1987).

Other factors, such as outdoor pollution levels and exposure to parental smoking, might have contributed to overall effects observed by Melia et al. and Speizer et al., as cited in DOE (1987). There is, however, no evidence in their studies to suggest that these factors differ for children living in electric versus gas-stove buildings (DOE 1987).

While animal studies provide some evidence that NO_2 impairs respiratory defense mechanisms, this evidence comes from studies conducted at considerably higher NO_2 exposure levels than those experienced in buildings with gas stoves (DOE 1987).

Speizer et al. (1980), as cited in DOE (1987), hypothesized that repeated peak values are probably the most important factor causing the effects observed in buildings with gas stoves. Their opinion is based, in part, on two facts: 1) there are no intermittent short-term (1/2 to 2 hours) NO_2 peak concentrations in buildings with electric stoves, and 2) long-term (24 hours or longer) concentrations in buildings with gas stoves are not markedly higher than in buildings with electric stoves (DOE 1987).

Effects of acute exposure to high NO_2 concentrations in humans, as summarized by the National Research Council (1977), as cited in DOE (1987), are shown in Table 2.16. The data in this table suggest a fatality threshold between 50 and 150 ppm, but exposure durations are not listed. A National

TABLE 2.16.   Human Effects of Acute Exposure to High NO_2 Concentrations

NO_2 Concentration mg/m^3	ppm	Clinical Effect	Time Between Exposure and Termination of Effect
940	500	Acute pulmonary edema (fatal)	Within 48 hours
564	300	Bronchopneumonia (fatal)	2 to 10 days
282	150	Bronchiolitis fibrosa obliterans (fatal)	3 to 5 weeks
94	50	Bronchiolitis, focal pneumonitis (recovery)	6 to 8 weeks
47	25	Bronchitis, bronchopneumonia (recovery)	6 to 8 weeks

SOURCE: National Research Council 1977b, p. 269; as cited in DOE, 1987.

Research Council summary (1977), as cited in DOE (1987), of human responses to short-term NO_2 exposures without the presence of other air pollutants is shown in Table 2.17. The results of controlled exposure of mice, guinea pigs, hamsters, cats, and humans to NO_x, as reported by several investigators, are shown in Table 2.18 (DOE 1987).

Based on the best available information, EPA (1982a), as cited in DOE (1987), has recommended the following ambient air quality standards for NO_2:

- A 1-hour average NO_2 standard at some level below 0.5 ppm, or at the range of 0.15 ppm to 0.30 ppm, which would have to be met for a specified number of days in the calendar year.

- An annual standard in the range of 0.05 to 0.08 ppm as an alternative to the short-term standard.

An annual standard in the range of 0.05 to 0.08 ppm would appear to provide adequate protection against the potential and uncertain health effects that may be associated with exposure to short-term NO_2 levels. Such a standard could be used as a surrogate for a short-term standard. In addition, an

TABLE 2.17.   Summary of Human Responses to Short-Term NO_2 Exposures Alone[a]

Effect	NO_2 Concentration mg/m³	ppm	Time to Effect
Odor threshold	0.23	0.12	Immediate
Threshold for dark adaptation	0.14	0.075	Not reported
	0.50	0.26	Not reported
Increased airway resistance	1.3 to 3.8	0.7 to 2.0	20 minutes[b]
	3.0 to 3.8	1.6 to 2.0	15 minutes[c]
	2.8	1.5	45 minutes[c]
	3.8	2.0	45 minutes[d]
	5.6	3.0	45 minutes[e]
	7.5 to 9.4	4.0 to 5.0	40 minutes[f]
	9.4	5.0	15 minutes
	11.3 to 75.2	6.0 to 40.0	5 minutes
	13.2 to 31.8	7.0 to 17.0	10 minutes[g]
Decreased pulmonary diffusing capacity	7.5 to 9.4	4.0 to 5.0	15 minutes
Increased alveolar arterial pO_2	9.4	5.0	25 minutes[h]
No change in sputum histamine concentration	0.9 to 6.6	0.5 to 3.0	45 minutes

(a) Reprinted from National Research Council 1977; as cited in DOE, 1987.
(b) Exposure lasted 10 minutes. Effect on flow resistance was observed 10 minutes after termination of exposure.
(c) Effect was produced at this concentration when normal subjects and those with chronic respiratory disease exercised during exposure.
(d) Effect occurred at rest in subjects with chronic respiratory disease.
(e) Effect occurred at rest in normal subjects.
(f) Exposure lasted 10 minutes. Maximal effect on flow resistance was observed 30 minutes later.
(g) Also failed to find increased flow resistance over the range of NO_2 exposures from 5.1 to 30.1 mg/m³ (2.7 to 16.0 ppm).
(h) Effect occurred 10 minutes after termination of 15-minute exposure.

Note: NO_2 TLV-TWA = 3 ppm or 6 mg/m³; TLV-STEL = 5 ppm or 10 mg/m³ (ACGIH 1983). EPA-recommended 1-hour average outdoor NO_2 standard <0.5 ppm or at the range of 0.15 to 0.30 ppm, or annual standard in the range of 0.05 to 0.08 ppm (EPA 1982).

TABLE 2.18. Controlled Exposure to NO_2

Species: Mouse (6- to 8-weeks old)
Exposure: NO_2, 10 ppm, 2 hours/day, 5 days/week up to 30 weeks

Health Effects Observed: Lung damage, suppressed immune function with chronic
                exposure, enhanced immune reactivity with shorter exposures.

References: Holt et al. 1979

Species: Mouse (6- to 8-weeks old)
Exposure: NO, 10 ppm

Health Effects Observed: Paraseptal emphysema, suppressed immune function
                with chronic exposure, enhanced immune reactivity with
                shorter exposures.

References: Holt et al. 1979

Species: Mouse
Exposure: NO_2, 0.5 to 28 ppm, 6 months to 1 year

Health Effects Observed: Mortality after Streptococcus pyogenes, challenge
                mortality increased with increasing dose and exposure time.

References: Larsen et al. 1979

Species: Guinea pig
Exposure: NO_x, 1 ppm, 6 months

Health Effects Observed: Disturbed glycolysis, enhanced catabolic processes
                in brain, inhibited respiration, decreased brain
                aminotransferase activity, morphologic alterations in blood
                vessels.

References: Kosmider et al. 1974

Species: Human (asthma, N = 13; bronchitis, N = 7)
Exposure: NO_2, 0.5 ppm

Health Effects Observed: Lightness in chest, burning of eyes, headache, or
                dyspnea; pulmonary-function changes; nasal discharge

References: Kerr et al. 1979

TABLE 2.18. (contd)

Species: Human (asthma, N = 20)
Exposure: NO_2, 0.1 to 0.2 ppm

Health Effects Observed: Increased bronchoconstriction[a]

References: Orehek et al. 1979

Species: Cat
Exposure: NO_2, 80 ppm, 3 hours

Health Effects Observed: Diffuse alveolar damage

References: Langloss et al. 1977

Species: Guinea pig
Exposure: NO_2, 0.506 ppm; NO, 0.05 ppm; 122 days

Health Effects Observed: In lungs: decreased phosphatidylethanolamine,
                sphingomyelin, phosphatidylserine, phosphatidic acid,
                phosphatidylglycerol 3-phosphate; increased
                lysophosphatidylethanolamine.

References: Trzeciak et al. 1977

Species: Mouse
Exposure: NO_2, 1.5 to 5.0 ppm, 3 hours

Health Effects Observed: Mortality in mice challenged with Streptococcus
                aerosol significantly increased at 2.0 ppm and above.

References: Ehrlich et al. 1977

Species: Mouse
Exposure: NO_2, 0.5 ppm; 10, 12, and 14 days

Health Effects Observed: Average protein content of lungs significantly
                higher.

References: Sherwin and Layfield 1976

TABLE 2.18.   (contd)

Species: Hamster
Exposure: NO_2, 30 ppm, 3 weeks

Health Effects Observed: Loss of body weight, increased dry lung weight,
                decreased lung elastin and collagen[b]

References: Kleinerman and Ip 1979

(a) Carbachol provocation.
(b) Elastin and collagen later returned to normal.

SOURCE: National Research Council 1981, p. 358; as cited in DOE, 1987.

annual standard would provide some, although unquantifiable, protection
against possible adverse health effects from long-term exposure (DOE 1987).

The lack of scientifically demonstrated health effects in humans from
NO_2 exposure in concentrations below 0.5 ppm could be interpreted to mean that
there is no need for an NO_2 National Ambient Air Quality Standards (NAAQS).
However, such an interpretation would ignore the cumulative evidence from
controlled animal and human exposure studies, and community indoor studies,
that strongly suggest that NO_2 can cause adverse health effects in sensitive
population groups exposed to NO_2 levels at or near existing ambient levels
(DOE 1987).

2.4 FIBERS

Mineral fibers include naturally occurring fibers, predominantly
asbestos, and synthetic fibers, predominantly fiberglass and mineral (rock)
wool, together with several other lesser-known species. Asbestos is a generic
term referring to hydrated magnesium (with the exception of crocidolite)
silicate fibers characterized by flexibility, strength, and resistance to fire
and chemicals. These desirable characteristics have resulted in numerous
industrial applications of asbestos, predominantly chrysotile. These applications include, but are not restricted to, thermal and acoustic insulation in
buildings and ships; additives to construction and building (sheets, pipes,

panels, cement, plaster, etc.), friction (brake linings and clutches), coating, roofing and flooring materials; filters and gaskets; and textiles (e.g., fire-resistant clothing). Table 2.19 lists several types of fibers potentially encountered in public buildings (DOE 1987).

TABLE 2.19. Types of Fibers Potentially Encountered in Public Buildings

Naturally Occurring Fibers

    Asbestos

        Serpentine

            (1) Chrysotile - $Mg_3(Si_2O_3)(OH)_4$

        Amphiboles

            (1) Actinolite - $Ca_2(Mg,Fe)_5(Si_8O_{22})(OH)_2$
            (2) Amosite - $(Fe,Mg)_7(Si_8O_{22})(OH)_2$
            (3) Anthophyllite - $(g,Fe)_7(Si_8O_{22})(OH)_2$
            (4) Crocidolite - $Na_2FeII_3FeIII_2(Si_8O_{22})(OH)_2$
            (5) Tremolite - $Ca_2Mg_5(Si_8O_{22})(OH)_2$

    Nonasbestos Fibers

        Attapulgite (palygorskite)

        Sepiolite (including meerschaum)

        Noncommercial natural mineral fibers

Synthetic Fibers

    Vitreous Fibers

        (1) Fibrous glass
        (2) Mineral wool (rock wool, slag wool)
        (3) Ceramic fibers

    Other Synthetic Fibers

        (1) Carbon fibers
        (2) Miscellaneous others

2.4.1 <u>Asbestos</u>

Asbestos is the name applied to a group of six different minerals that occur naturally in the environment. The most common mineral type is white, but others may be blue, gray or brown. These minerals are made up of long, thin fibers that are somewhat similar to fiberglass. Asbestos fibers are very strong and are resistant to heat and chemicals. Because the fibers are so resistent to chemicals, they are also very stable in the environment; they do not evaporate into air or dissolve in water, and they are not broken down over time. Asbestos has a Hazard Rating of 3 (Sax and Lewis 1989).

2.4.1.1 <u>Summary</u>

The present EPA carcinogen assessment summary for asbestos may change in the near future pending the outcome of further review now being conducted by the CRAVE Work Group. The weight-of-evidence classification for asbestos is "A; human carcinogen." The basis is observation of increased mortality and incidence of lung cancer, mesotheliomas and gastrointestinal cancer in exposed workers. Animals studies by inhalation in two strains of rats showed similar findings for lung cancer and mesotheliomas. Animal evidence for carcinogenicity via ingestion is limited (male rats fed intermediate-range chrysotile fibers; i.e., greater than 10 μm length, developed benign polyps), and epidemiological data in this regard are inadequate (IRIS 1990).

The Inhalation Slope Factor for asbestos is 2.3E-1/fibers/mL. The Inhalation Unit Risk estimate for asbestos is 2.3E-1/fibers/mL. The extrapolation method used was additive risk of lung cancer and mesothelioma, using relative risk model for lung cancer and absolute risk model for mesothelioma.

2.4.1.2 <u>Health Effects</u>

<u>Human Carcinogen</u> - The present EPA carcinogen assessment summary for asbestos may change in the near future pending the outcome of further review now being conducted by the CRAVE Work Group. The weight-of-evidence classification for asbestos is "A; human carcinogen." The basis is observation of increased mortality and incidence of lung cancer, mesotheliomas and gastrointestinal cancer in exposed workers. Animals studies by inhalation in two strains of rats showed similar findings for lung cancer and mesotheliomas.

Animal evidence for carcinogenicity via ingestion is limited (male rats fed intermediate-range chrysotile fibers; i.e., greater than 10 μm length, developed benign polyps), and epidemiological data in this regard are inadequate (IRIS 1990).

Human Carcinogenicity Data - The human carcinogenicity data are considered sufficient. Numerous epidemiologic studies have reported an increased incidence of deaths due to cancer, primarily lung cancer and mesotheliomas associated with exposure to inhaled asbestos. Among 170 asbestos insulation workers in North Ireland followed for up to 26 years, an increased incidence of death was seen from all cancers (SMR=390), cancers of the lower respiratory tract and pleura (SMR=1760) (Elmes and Simpson 1971, as cited in IRIS (1990) and mesothelioma (7 cases). Exposure was not quantified.

Selikoff (1976) reported 59 cases of lung cancer and 31 cases of mesothelioma among 1249 asbestos insulation workers followed prospectively for 11 years. Exposure was not quantified. A retrospective cohort mortality study (Selikoff et al. 1979, as cited in IRIS 1990) of 17,800 U.S. and Canadian asbestos insulation workers for a 10-year period using best available information (autopsy, surgical, clinical) reported an increased incidence of cancer at all sites (319.7 expected vs. 995 observed, SMR=311) and cancer of the lung (105.6 expected vs. 486 observed, SMR=460). A modest increase in deaths from gastrointestinal cancer was reported with 175 deaths from mesothelioma (none expected). Years of exposure ranged from less than 10 to greater than or equal to 45. Levels of exposure were not quantified. In other epidemiologic studies, the increase for lung and pleural cancers has ranged from a low of 1.9 times the expected rate, in asbestos factory workers in England (Peto et al. 1977; as cited in IRIS 1990), to a high of 28 times the expected rate, in female asbestos textile workers in England (Newhouse et al. 1972, as cited in IRIS 1990). Other occupational studies have demonstrated asbestos exposure-related increases in lung cancer and mesothelioma in several industries including textile manufacturing, friction products manufacture, asbestos cement products, and in the mining and milling of asbestos. The studies used for the inhalation quantitative estimate of risk are listed in Table 2.20.

TABLE 2.20.   Asbestos Dose-Response Data for Carcinogenicity,
Inhalation Exposure

Human Data Occupational Group	Fiber Type	Exposure (fiber yr/mL)	Reported Average % Increase in Cancer/ fiber-yr/mL	Reference (as cited in IRIS 1990)
Lung Cancer:				
Textile Prod.	Predominantly Chrysotile	44	2.8	Dement et al. (1983)
Textile Prod.	Chrysotile	31	2.5	McDonald et al. (1983A)
Textile Prod.	Chrysotile	200	1.1	Peto (1980)
Textile Prod.	Chrysotile	51	1.4	McDonald et al. (1983b)
Friction Prod.	Chrysotile	32	0.058	Berry and Newhouse (1983)
Friction Prod.	Chrysotile	31	0.010	McDonald et al. (1984)
Insulation Products	Amosite	67	4.3	Seidman (1984)
Insulation Workers	Mixed (Chrysotile Crocidolite and Amosite)	300	0.75	Selikoff et al. (1979)
Asbestos Products		374	0.49	Henderson and Enterline (1979)
Cement Products		89 112	0.53 6.7	Weill et al. (1979) Finkelstein (1983)
Mesothelioma:				
Insulation Workers	Mixed	375	1.5E-6	Selikof et al. (1979); Peto et al. (1982)
Insulation	Amosite	400	1.0E-6	Seidman et al. (1979)
Textile Prod. Manufacturer	Chrysotile	67	3.2E-6	Peto (1980); Peto et al. (1982)
Cement Prod.	Mixed	108	1.2E-5	Finkelstein (1983)

A case-control study (Newhouse and Thompson 1965, as cited in IRIS 1990) of 83 patients with mesothelioma reported 52.6% had occupational exposure to asbestos or lived with asbestos workers compared with 11.8% of the controls. Of the remaining subjects, 30.6% of the mesothelioma cases lived within one-half mile of an asbestos factory compared with 7.6% of the controls (IRIS 1990).

The occurrence of pleural mesothelioma has been associated with the presence of asbestos fibers in water, fields and streets in a region of Turkey with very high environmental levels of naturally-occurring asbestos (Baris et al. 1979, as cited in IRIS 1990).

Kanarek et al. (1980), as cited in IRIS (1990), conducted an ecologic study of cancer deaths in 722 census tracts in the San Francisco Bay area, using cancer incidence data from the period of 1969-1971. Chrysotile asbestos concentrations in drinking water ranged from nondeductible to 3.6E+7 fibers/L. Statistically significant dose-related trends were reported for lung and peritoneal cancer in white males and for gall bladder, pancreatic and peritoneal cancer in white females. Weaker correlations were reported between asbestos levels and female esophageal, pleural and kidney cancer, and stomach cancer in both sexes. In an extension of this study, Conforti et al. (1981), as cited in IRIS (1990), included cancer incidence data from the period of 1969-1974. Statistically significant positive associations were found between asbestos concentration and cancer of the digestive organs in white females, cancer of the digestive tract in white males and esophageal and pancreatic and stomach cancer in both sexes. These associations appeared to be independent of socioeconomic status and occupational exposure to asbestos (IRIS 1990).

Marsh (1983), as cited in IRIS (1990), reviewed eight independent ecologic studies of asbestos in drinking water carried out in five geographic areas. It was concluded that even though one or more studies found an association between asbestos in water and cancer mortality (or incidence) from neoplasms of various organs, no individual study or aggregation of studies exists that would establish risk levels from ingested asbestos. Factors confounding the results of these studies include the possible underestimates of occupational exposure to asbestos and the possible misclassification of peritoneal mesothelioma as GI cancer (IRIS 1990).

Polissar et al. (1984), as cited in IRIS (1990), carried out a case-control study which included better control for confounding variables at the individual level. The authors concluded that there was no convincing evidence for increased cancer risk from asbestos ingestion. At the present time, an important limitation of both the case-control and the ecologic studies is the short follow-up time relative to the long latent period for the appearance of tumors from asbestos exposure (IRIS 1990).

Animal Carcinogenicity Data - The animal carcinogenicity data are considered sufficient. There have been about 20 animal asbestos bioassays. The results of some of the more significant of these bioassays are provided in IRIS (1990). However, because the human carcinogenicity data is also considered sufficient, that information is not presented in this report.

Quantitative Estimate of Carcinogenic Risk - The Inhalation Slope Factor for asbestos is 2.3E-1/fibers/mL. The Inhalation Unit Risk estimate for asbestos is 2.3E-1/fibers/mL. The extrapolation method used was additive risk of lung cancer and mesothelioma, using relative risk model for lung cancer and absolute risk model for mesothelioma. The air concentrations at specified risk levels are as follows:

Risk Level	Concentration
E-4 (1 in 10,000)	4E-4 fibers/mL
E-5 (1 in 100,000)	4E-5 fibers/mL
E-6 (1 in 1,000,000)	4E-6 fibers/mL

Inhalation Carcinogenicity Exposure Comments - Risks have been calculated for males and females according to smoking habits for a variety of exposure scenarios (EPA 1986, as cited in IRIS 1990). The unit risk value is calculated for the additive combined risk of lung cancer and mesothelioma, and is calculated as a composite value for males and females. The epidemiological data show that cigarette smoking and asbestos exposure interact synergistically for production of lung cancer and do not interact with regard to mesothelioma. The unit risk value is based on risks calculated using U.S. general population cancer rates and mortality patterns without consideration of smoking habits. The risks associated with occupational exposure were adjusted to continuous exposure by applying a factor of 140 m^3 per 50 m^3 based on the assumption of 20 m^3/day for total ventilation and 10 m^3/8-hour workday in the occupational setting (IRIS 1990).

The unit risk is based on fiber counts made by phase contrast microscopy (PCM) and should not be applied directly to measurements by other analytical techniques. The unit risk uses PCM fibers because the measurements made in the occupational environment use this method. Many environmental monitoring measurements are reported in terms of fiber counts or mass as determined by transmission electron microscopy (TEM). PCM detects only fibers longer than 5 μm and greater than 0.4 μm in diameter, but TEM can detect much smaller fibers. TEM mass units are derived from TEM fiber counts. The correlation between PCM fiber counts and TEM mass measurements is very poor. Six data sets that include both measurements show a conversion between TEM mass and PCM fiber count that range from 5-150 (μg/m^3)/fibers/mL). The geometric mean of these results, 30 (μg/m^3)/fibers/mL), was adopted as a conversion factor (EPA 1986, as cited in IRIS 1990), but it should be realized that this value is highly uncertain. Likewise, the correlation between PCM and TEM fiber counts is very uncertain and no generally applicable conversion factor exists for these two measurements (IRIS 1990).

In some cases, TEM results are reported as numbers of fibers less than 5 μm long and of fibers longer than 5 μm. Comparison of PCM fiber counts and TEM counts of fibers greater than 5 μm show that the fraction of fibers detected by TEM that are also greater than 0.4μm in diameter (and detectable by PCM) varies from 22-53% (EPA 1986, as cited in IRIS 1990).

It should be understood that while TEM can be specific for asbestos, PCM is a nonspecific technique and will measure any fibrous material. Measurements by PCM that are made in conditions where other types of fibers may be present may not be reliable (IRIS 1990).

Some evidence suggests that different types of asbestos fibers vary in carcinogenic potency relative to one another and site specificity. It appears, for example, that the risk of mesothelioma is greater with exposure to crocidolite than with amosite or chrysotile exposure alone. This evidence is limited by the lack of information on fiber exposure by mineral type. Other data indicate that differences in fiber size distribution and other process differences may contribute at least as much to the observed variation in risk as does the fiber type itself (IRIS 1990).

The unit risk should not be used if the air concentration exceeds $4E-2$ $\mu g/m^3$, because above this concentration the slope factor may differ from that stated (IRIS 1990).

A large number of studies of occupationally exposed workers have conclusively demonstrated the relationship between asbestos exposure and lung cancer or mesothelioma. These results have been corroborated by animal studies using adequate numbers of animals. The quantitative estimate is limited by uncertainty in the exposure estimates, which results from a lack of data on early exposure in the occupational studies and the uncertainty of conversions between various analytical measurements for asbestos (IRIS 1990).

2.4.2 Synthetic Fibers

Biological effects of synthetic fibers have been reviewed by Hill (1977), Wagner et al. (1980), and Boatman et al. (1983), all as cited in DOE (1987). Some, but not all, of the animal studies show that synthetic fibers can result in pulmonary disease (Hill 1977; Bayliss et al. 1976; Stanton et al. 1977; Wright and Kuschner 1977; Boatman et al. 1983, all as cited in DOE 1987). Bayliss et al. (1976), as cited in DOE (1987), noted a significant excess of nonmalignant respiratory diseases in workers exposed to glass fibers. However, the evidence is not considered conclusive because of the number of studies showing no effects in humans or animals (Boatman et al. 1983, as cited in DOE 1987).

No epidemiological evidence exists at this time to link any of the synthetic fiber substitutes for asbestos to lung cancer. However, the use of these materials is relatively new. If they are indeed carcinogenic, and if cancers resulting from such exposures had latency periods similar to those induced by asbestos, these cancers would not manifest themselves epidemiologically until after the turn of the century (DOE 1987).

Lack of convincing evidence and reliable information at present does not permit a meaningful risk assessment of exposure to synthetic fibers in enclosed spaces (DOE 1987).

2.5 BIOGENIC PARTICLES

Biogenic indoor air pollutants include viruses; bacteria; fungi; algae; pollens and other plant-derived materials; protozoa; helminths; arthropods (particularly mites) and insects, their excretions and body fragments; bird feathers; dander of dogs and cats; and human epidermal scales. The dust in public buildings contain varying proportions of these materials. Some of these biogenic air pollutants originate outdoors and invade the buildings from there (e.g., pollens). Others originally migrated from outdoors but subsequently established themselves in the indoor environment (e.g., bacteria, fungi, arthropods, insects). Still others of them are generated indoors (e.g., human and pet danders).

2.5.1 Summary

In contrast to most of the chemical air pollutants (which may be toxic, carcinogenic or both), biogenic air contaminants can cause infectious diseases (e.g., colds, influenza, measles, chicken pox, smallpox, tuberculosis, Legionnaires' disease, pneumonia and several fungal diseases of the lung such as histoplasmosis, and aspergillosis). Biogenic air contaminants can cause allergic reactions such as allergic or vasomotor rhinitis (hay fever), asthma, and hypersensitivity pneumonitis (extrinsic allergic bronchioloalveolitis). Tissue changes caused by acute allergic reactions are reversible. However, if exposure to the offending allergen turns chronic, reversibility diminishes and eventually irreversible lesions will develop, often in the form of pulmonary fibrosis (lung scarring) (DOE 1987).

Little is known about indoor air concentrations of biogenic particles in public buildings and the effect of environmental changes on those concentrations. Systematic standardized and coordinated measurements would be required to provide an essential information base for developing guidelines for exposures to biogenic indoor air pollutants. The problem is complicated by the fact that, owing to the great diversity in physicochemical and biological characteristics of biogenic particles, no single sampling or sample processing technique can cover the entire spectrum of biogenic particles. Therefore, no meaningful risk assessment of exposure to biogenic indoor air pollutants in public buildings can be made at this time.

2.5.2 Health Effects

General Health Effects - While harmful chemical air contaminants are toxic/carcinogenic, biogenic air pollutants usually show very low, if any, toxic/carcinogenic effects. Instead, they can cause infections or allergic responses, or both.

Aerial transmission of pathogenic viruses indoors can cause diseases such as upper respiratory infections (colds, influenza), lung disease (e.g., psittacosis), rubella (measles), varicella (chicken pox), and variola (smallpox) (Leedom and Loosli 1979; Zeterberg 1973; McLean et al. 1967, all as cited in DOE 1987). Aerial transmission of pathogenic bacteria can result in bacterial upper respiratory infections and lung disease such as tuberculosis, Legionaires' disease, and pneumonia. Kelsen and McGuckin (1980), as cited in DOE (1987), for example, observed a statistically significant relationship between airborne microbial counts (CFP) and respiratory attack rates in patients with nosomial pneumonia. Several genera of fungi are pathogenic when inhaled by humans, among them the highly infectious Histoplasma and Coccidioides, as well as Aspergillus, Blastomyces, Cryptococcus, Candida albicans, and dermatophytes (EPA 1987).

Colds and influenza alone, mostly transmitted by airborne infectious agents in the indoor environment, cause an enormous burden for the national economy in terms of loss of productivity and exposures. They account for more than 50% of all acute conditions with an incidence of about one per person per year (DHEW 1975, as cited in DOE 1987). Afflicted patients are incapacitated or restricted in their activities for an average of 4.5 days. Respiratory conditions cause more loss of time from work or school than any other disease (DOE 1987).

Allergic reactions generally occur on the skin or in the mucous membranes of the respiratory tract. An allergic response to inhaled biogenic

agents manifests itself generally as a local inflammatory reaction at the site
of particle impingement. An estimated 15% of Americans suffer from reactions
to airborne allergens. If the reaction takes place mainly in the nasal area,
where larger inhaled particles such as pollen grains are retained, the
condition is referred to as allergic or vasomotor rhinitis, more popularly
know as hay fever. Allergic rhinitis is characterized by edema, swelling,
vasodilation of nasal mucosa, mucus hypersecretion, nasal discharge, and
congestion of the nasal airways.

Asthma refers to a partial, temporary narrowing of the bronchi from
spasm of the smooth bronchial muscles, edema of the bronchial mucosa, mucus
accumulation, or a combination of these factors. The fungus _Aspergillus
fumigatus_, when inhaled, can affect bronchi and the alveolar region by a
relatively rare condition called allergic bronchopulmonary aspergillosis. The
disease is characterized by recurrent episodes of temporary shadowing on chest
X-rays and in increased eosinophilic blood cells, often associated with asthma
attacks. Histologically, eosinophils accumulate in the alveolar spaces, often
leading to consolidation. As the disease progresses, the reversibility of
tissue changes decreases; eventually chronic bronchial stenosis, bron-
chiectasis, and pulmonary fibrosis might develop.

Hypersensitivity pneumonitis, also called extrinsic allergic bronchioloalveolitis, refers to an allergic reaction in the peripheral bronchioles and
alveoli between inhaled biogenic agents and circulating antibodies and
sensitized lymphocytes. In its acute stage, the disease is characterized by
infiltration of mononuclear cells into, and thickening of, alveolar septa and
bronchioles, often accompanied by the formation of epithelial and giant-cell
granulomas; when the disease progresses from continued allergen exposure,
fibrosis can develop. Reaction sites and allergic diseases of the respiratory
system are summarized in Table 2.21 (DOE 1987).

While the mechanism of sensitization in allergic individuals is still
unclear, it is well known that a single pollen can provoke a severe allergic
reaction in sensitized persons. Allergens maintain their allergenic effects,
independent of the viability of the microbe involved (if any), for considerable periods of time until the proteins of the allergen are denatured.

Rimington et al. (1947), as cited in DOE (1987), for example, reported that house dust remained allergenic until all amino acids had been hydrolyzed (DOE 1987).

Among the various indoor air pollutants, biogenic particles are probably the most complex and least investigated materials. Reliable measurements of airborne biogenic particle concentrations in residences are lacking and the effects of controlled variables on these concentrations is unknown. The scarce information presently available on this subject is insufficient for developing guidelines for human exposure to biogenic indoor air contaminants.

Hypersensitivity Pneumonitis and Humidifier Fever - Outbreaks of interstitial lung disease and febrile syndromes are among the best documented building-related diseases. In addition to outbreaks, our epidemiologic understanding has been supplemented by numerous case reports documenting illness resulting from exposure to allergens from home humidifiers, air coolers, car air conditioners, and saunas. Symptoms have varied, even within an outbreak, from acute recurrent pneumonias to insidious progression of

TABLE 2.21. Allergic Respiratory Diseases - Reaction Sites, Diseases, and Immunologic Mechanisms

Reaction Site	Disease	Immunologic Mechanism
Nose	Allergic rhinitis	IgE
Airways	Allergic rhinitis	IgE
Airways and alveolar spaces	Asthma with pulmonary eosinophilia, allergic bronchopulmonary aspergillosis	IgE, IgG, immune complexes
Alveolar walls peripheral bronchioles	Hypersensitivity pneumonitis (extrinsic allergic bronchioloalviolitis)	IgG, immune and complexes, sensitized T-lymphocytes

SOURCE: National Research Council (1981), p. 398, as cited in DOE (1987).

cough, shortness of breath, and fatigue that the patient does not attribute to indoor air exposure. In addition to the pulmonary diseases, recurrent outbreaks of fever, leukocytosis, chills, muscle aches, and malaise, without prominent pulmonary symptoms or signs, are part of this disease spectrum. Individual outbreaks have had unusual association symptoms such as polyuria, nausea, or headache, follicular conjunctivitis, and diarrhea, in addition to respiratory complaints. Five occupational outbreaks have had a pattern of recurrence on Monday evenings or on the evening of the first day back to work, reminiscent of byssinosis and metal fume fever. On the other hand, symptoms worsened toward the end of the work week in one case, and no association with day of the work week is common (Walsh et al. 1984).

Attack rates have varied from 1 to 71%. Various bacteria and fungi have been implicated in outbreaks and case reports, but pure strains of organisms found in humidifier water or other sources are sometimes ineffective in bronchial challenge testing or in precipitin tests with sera from affected individuals. Protozoa may play a role in some of these outbreaks and may require special methods for detection in environmental samples (Walsh et al. 1984).

Infections - Many infectious diseases can be transmitted via indoor air. These infectious diseases include Legionnaires' disease, Pontiac fever, and Q fever.

Legionnaires' disease, a bacterial pneumonia, was first described in the context of an epidemic of 182 cases (20 fatal) among Legionnaires attending a convention in 1976 at the Bellevue Stratford Hotel in Philadelphia. Since the identification of the bacterial agent, Legionella pneumophila, many building-associated epidemics have been recognized, both retrospectively and prospectively. The multisystem illness involves the GI tract, kidney, central nervous system, lungs. From 1% to 7% of persons exposed become ill after an average incubation period of 5 to 6 days. The case-fatality rate is approximately 15%. Epidemics have been associated with aerosols from cooling towers and evaporative condensers and with dusts from landscaping and construction requiring soil excavation. In addition, the organism has been cultured from shower heads in hospitals and hotels associated with cases of Legionnaires' disease. An epidemic of seven cases associated with an inn with a whirlpool occurred over a 15-month period from May 1980 to August 1981. Outbreaks of Legionnaires' disease are no longer etiologic puzzles to public health investigators, although many questions remain concerning Legionella ecology and infectivity (Walsh et al. 1984).

Pontiac fever was first described as a building-related epidemic of 144 cases in a county health department in Pontiac, Michigan in 1968. The air conditioning system was contaminated with L. pneumophila and was the mode of dissemination. Two subsequent outbreaks have been reported: one among 10 men who spent 9 hours cleaning a steam turbine condenser with compressed air and one in an automobile assembly plant thought to be caused by an aerosol of a contaminated oil-based coolant. In contrast to Legionnaires' disease, Pontiac fever is a 2- to 5-day illness characterized by fever, chills, headache, and myalgia. Cough, sore throat, chest pain, nausea, and diarrhea may be present, but in lower prevalence. The attack rate among exposed persons is character- istically 95 to 100%, with a mean incubation period of about 36 hours (Walsh et al. 1984).

Although Pontiac fever outbreaks are diagnosed by finding L. pneumophila organisms in environmental samples and diagnostic rises in antibody titer to the organism in convalescent-phase sera, no consensus exists regarding why Legionellae cause outbreaks of two distinct clinical syndromes. No convincing evidence supports the importance of dose, toxin production, or vital status of the inhaled organism. Rowbotham, as cited in Walsh (1984), has suggested that Pontiac fever represents a hypersensitivity pneumonitis from amoebae with a limited infection by L. pneumophila in most cases. Free-living amoebae are hosts to and victims of Legionella infection, depending on number and strain of each, and temperature. Inhalation of amoebae or amoebic vesicles might deliver many more Legionellae to the human host. Humidifier fever-type sensitization of amoebal antigens may interfere with the establishment of Legionella infection, or amoebic digestion of Legionellae may reduce the dose below that effective for clinical pneumonia. The recent epidemic of Legionellosis in the inn with a whirlpool is the first report of a Legionella source associated with sporadic cases of Legionnaires' disease over a 15-month

period and an epidemic of 34 cases of Pontiac fever in March 1981. Three
groups of Legionella were isolated from whirlpool water: L pneumophila,
serogroup 1 and 6, and L. dumoffi. Studies for amoebae and other protozoa
were not performed before the whirlpool was closed in August 1981 (Walsh et
al. 1984).

Q fever, a rickettsial illness caused by Coxiella burnetii, has caused
several building associated epidemics. Symptoms resemble influenza with
abrupt onset of high fever, chills, headache, and myalgia. In some outbreaks,
pneumonia has been a prominent manifestation, and cases of hepatitis and
endocarditis occur rarely. The reservoir for infection is infected sheep,
goats, and cattle, whose excreta and placental tissue can contaminate buildings in which these animals are housed. All of the outbreaks cited occurred
in university hospitals or research laboratory buildings in which the organism
was cultured; however, in each outbreak, people without direct contact with
animals or the organism became ill as a result of airborne transmission in the
building (Walsh et al. 1984).

2.6 OTHER POTENTIAL CONTAMINANTS

To assess health impacts from contaminants in the indoor air of public
buildings, contaminants that may affect human health must first be identified.
Hundreds of such biological and chemical contaminants exist. An EPA study
indicated at least 500 different volatile organic compounds have been
identified in the indoor air of buildings (EPA 1988b).

As stated in Section 1.4 in this report, we considered only contaminants
that were measured in indoor air of public buildings or those measured in test
situations that simulated indoor air quality situations, as reported in the
literature examined as a basis for selecting contaminants to study (see
Section 1.4 for list of literature references). Contaminants found in building materials, but not reported as being found in any of the indoor air
studies reviewed, were not included in this study. Therefore, there are many
other potential contaminants that may be of concern in assessing the risk to
human health from a public building.

For example, some of the contaminants identified (in the literature reviewed) in building materials but not found in the building studies examined include cumene, chlorobenzene, nonane, iso-octane, iso-nonane, iso-decane, tetrachloroethane, mesitylene, n-hexane, 3-carene, propanol, 2-butanone, acetone, diethylbenzene, hexanol, d-limonene, vinyl chloride, butanol, n-heptane, n-heptene-1, 3-pentanol, and ethylacetate.

The additional and very complex problem of complex chemical mixtures must also be addressed to fully understand and predict the risk from chemicals found in the indoor environment of public buildings. Though the health effect studies of the individual chemical compounds found in indoor air is extremely useful in understanding and predicting the risks from these chemicals, there are serious limitations in extrapolating the results from single chemical studies to the actual environmental exposure of humans, which is that of complex mixtures of these various chemical compounds. The toxicological study of complex mixture sets representing actual exposure in typical public building situations should be performed to more accurately examine the combined or synergistic effect of the actual chemical mixture set to which the public is being exposed.

3. Polychlorinated Biphenyls

Polychlorinated biphenyls (PCBs) are a series of technical mixtures consisting of many isomers and compounds that vary from mobile oily liquids to white crystalline solids and hard noncrystalline resins. The technical products vary in composition, in the degree of chlorination and possibly according to batch. With a Hazard Rating of 3 (Sax and Lewis, 1989), the molecular weight of PCBs varies from 189-399 g/mole. They have a flash point of 200°C and a vapor pressure of 7.7E-5 mm Hg (ATSDR 1987).

3.1 SUMMARY

In humans exposed to PCBs, reported adverse effects include chloracne (a long-lasting, disfiguring skin disease), impairment of liver function, a variety of neurobehavioral and affective symptoms, menstrual disorders, minor birth abnormalities, and probably increased incidence of cancer. Animals experimentally exposed to PCBs have shown most of the same symptoms, as well as impaired reproduction; pathological changes in the liver, stomach, skin, and other organs; and suppression of immunological functions. PCBs are carcinogenic in rats and mice and, in appropriate circumstances, enhance the effects of other carcinogens. Reproductive and neurobiological effects of PCBs have been reported in rhesus monkeys at the lowest dose level tested, 11 g/kg body weight per day over a period of several months (ATSDR 1987).

The weight-of-evidence classification for PCBs is "B2; probable human carcinogen." The basis is hepatocellular carcinomas in three strains of rats and two strains of mice and inadequate, yet suggestive, evidence of excess risk of liver cancer in humans by ingestion and inhalation or dermal contact (IRIS 1990).

The Oral Slope Factor for PCBs is 7.7/mg/kg/day. The Drinking Water Unit Risk is 2.2E-4/μg/L. The extrapolation method used was the linear multistage procedure with extra risk.

263

3.2 HEALTH EFFECTS

The health effects discussed for PCBs focus on the inhalation route because this is the primary pathway for exposure of an individual to PCBs via indoor air in a public building. However, dermal exposure could also occur. Therefore, dermal route toxicity summaries are also provided. Oral route risk factors for PCBs are provided because such factors are only available for the oral route.

Lethality and Decreased Longevity (Inhalation) - Data regarding inhalation exposure levels that produce death in humans were not available. Exposure to near-saturation vapor concentrations of heated Aroclor 1242 (a PCB) (8.6 mg/m^3) 7 h/day, 5 day/wk for 24 days was not lethal for cats, rats, mice, rabbits, or guinea pigs (Treon et al., 1956, as cited in ATSDR 1987). This concentration represents a NOAEL for lethality for intermediate inhalation exposures. Figures 3.1, 3.2, 3.3 and 3.4 show effects of PCB-inhalation exposure, PCB-oral exposure, PCB-dermal exposure and levels of significant exposure for PCBs-inhalation, respectively. No data were available regarding lethality/decreased longevity of animals due to acute or chronic inhalation exposure to PCBs (ATSDR 1987).

Systemic/Target Organ Toxicity (Inhalation) - Oral toxicity studies in animals have established that the liver and cutaneous tissues are primary target organs of PCBs with increased serum levels of liver-associated enzymes and dermatologic effects such as chloracne and skin rashes. The results of some of these studies are equivocal, and exposure levels were not reported or were inadequately characterized. Furthermore, although inhalation is considered a major route of exposure, the contribution of dermal exposure to total occupational exposure is also significant (ATSDR 1987).

Fischbein et al. (1979, 1982, 1985), as cited in ATSDR (1987), reported data suggesting associations between serum levels of PCBs and (serum glutamic oxaloacetic transaminase (SGOT) levels and dermatologic effects in workers who had been exposed to 8-hour, time-weighted, average concentrations of Aroclors, primarily 1242 and 1254, ranging from 0.007 to 11.0 mg/m^3. Because of limitations of this study, these effects could be regarded as inconclusive and cannot be associated with specific exposure concentrations.

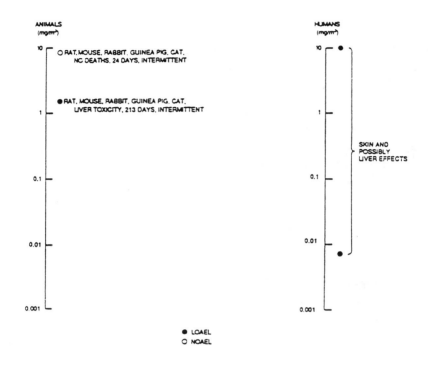

FIGURE 3.1. Effects of PCBs, Inhalation Exposure (ATSDR 1987)

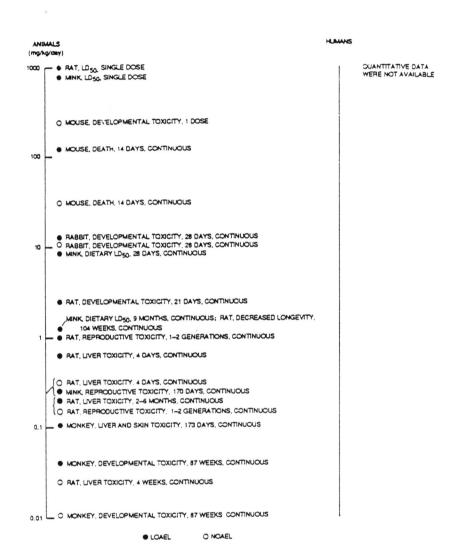

ANIMALS
(mg/kg/day)

HUMANS

QUANTITATIVE DATA
WERE NOT AVAILABLE

1000 — ● RAT, LD$_{50}$, SINGLE DOSE
● MINK, LD$_{50}$, SINGLE DOSE

○ MOUSE, DEVELOPMENTAL TOXICITY, 1 DOSE

● MOUSE, DEATH, 14 DAYS, CONTINUOUS

100 —

○ MOUSE, DEATH, 14 DAYS, CONTINUOUS

● RABBIT, DEVELOPMENTAL TOXICITY, 28 DAYS, CONTINUOUS
10 — ○ RABBIT, DEVELOPMENTAL TOXICITY, 28 DAYS, CONTINUOUS
● MINK, DIETARY LD$_{50}$, 28 DAYS, CONTINUOUS

● RAT, DEVELOPMENTAL TOXICITY, 21 DAYS, CONTINUOUS

MINK, DIETARY LD$_{50}$, 9 MONTHS, CONTINUOUS; RAT, DECREASED LONGEVITY,
● 104 WEEKS, CONTINUOUS
1 — ● RAT, REPRODUCTIVE TOXICITY, 1–2 GENERATIONS, CONTINUOUS

● RAT, LIVER TOXICITY, 4 DAYS, CONTINUOUS

○ RAT, LIVER TOXICITY, 4 DAYS, CONTINUOUS
● MINK, REPRODUCTIVE TOXICITY, 170 DAYS, CONTINUOUS
● RAT, LIVER TOXICITY, 2–6 MONTHS, CONTINUOUS
○ RAT, REPRODUCTIVE TOXICITY, 1–2 GENERATIONS, CONTINUOUS
0.1 — ● MONKEY, LIVER AND SKIN TOXICITY, 173 DAYS, CONTINUOUS

● MONKEY, DEVELOPMENTAL TOXICITY, 87 WEEKS, CONTINUOUS

○ RAT, LIVER TOXICITY, 4 WEEKS, CONTINUOUS

0.01 — ○ MONKEY, DEVELOPMENTAL TOXICITY, 87 WEEKS, CONTINUOUS

● LOAEL   ○ NOAEL

FIGURE 3.2. Effects of PCBs, Oral Exposure

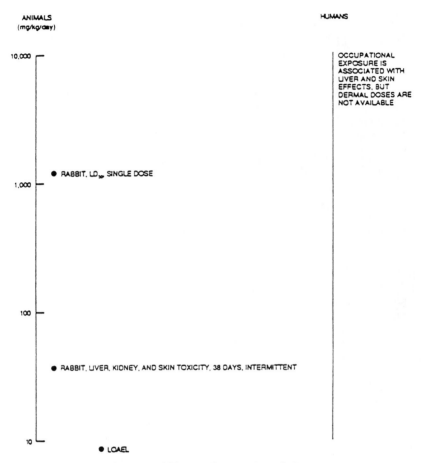

FIGURE 3.3. Effects of PCBs, Dermal Exposure

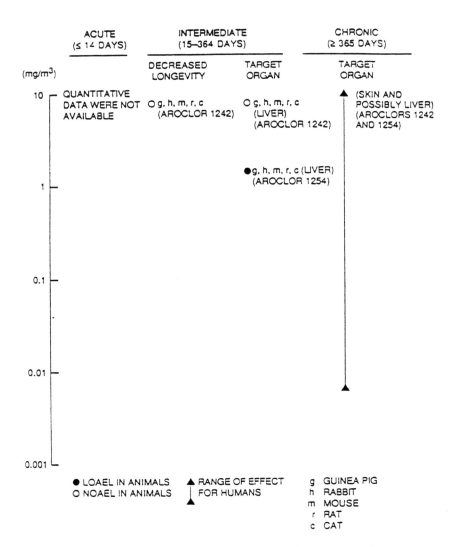

FIGURE 3.4. Levels of Significant Exposure for PCBs, Inhalation

It is, however, appropriate to plot the range of Aroclor concentrations from this study in Figures 3.1 and 3.4 because similar effects have been observed in other health surveys of PCB-exposed workers, information regarding human liver histopathology is lacking, and the liver and skin are unequivocal targets of PCB toxicity in animals. This concentration range is intended to approximate typical concentrations in occupational environments that may be associated with hepatic and dermatologic alterations (ATSDR 1987).

In only the animal inhalation study of PCBs, degenerative liver lesions, a frank effect, occurred in cats, rats, mice, rabbits, and guinea pigs that were exposed to 1.5 mg/m^3 Aroclor 1254 vapor for 7 h/day, 5 day/wk for 213 days (Treon et al., 1956, as cited in ATSDR 1987). This FEL is plotted on Figures 3.1 and 3.4. Histologic effects were not produced in those species exposed to Aroclor 1242 (1.9 mg/m^3, 7 h/day, 5 day/wk for 214 days; 8.6 mg/m^3, 7 h/day, 5 day/wk for 24 days). The higher NOAEL of 8.6 mg/m^3 for intermediate-duration inhalation exposure is plotted on Figure 3.4 since the FEL for Aroclor 1254 is lower than the NOAEL for Aroclor 1242, a minimal risk level cannot be derived for Aroclors as a class (ATSDR, 1987).

Developmental Toxicity (Inhalation) - Pertinent data regarding developmental effects of PCBs via inhalation exposure in animals were not located in the available literature. A report of slightly reduced birth weight and gestational age in infants born to mothers with occupational exposure to Aroclors (Taylor et al. 1984, as cited in ATSDR 1987) is inconclusive and lacks monitoring data.

Reproductive Toxicity (Inhalation) - Pertinent data regarding reproductive effects of PCBs via inhalation exposure in humans or animals are not available (ATSDR 1987).

Genotoxicity (Inhalation) - The PCBs have produced generally negative results in in vivo and in vitro genotoxicity assays (ATSDR 1987).

Human Carcinogenicity (Inhalation) - Occupational studies (Brown 1986, Bertazzi et al. 1987, as cited in ATSDR 1987) provide inadequate but suggestive evidence for carcinogenicity of PCBs by the inhalation route. Data regarding the carcinogenicity of inhaled PCBs in animals are not available (ATSDR 1987).

Lethality and Decreased Longevity (Dermal) - Human data are not available. Median lethal doses for single dermal applications of PCBs to rabbits ranged from less than 1,269 mg/kg for Aroclors 1242 and 1248 to less than 3,169 mg/kg for Aroclor 1221 (Fischbein, 1974, as cited in ATSDR 1987). Because only ranges of median lethal doses were reported, the lowest dose (1,269 mg/kg) is indicated on Figures 3.3 and 3.6 (ATSDR 1987).

Systemic/Target Organ Toxicity (Dermal) - The study of capacitor workers by Maroni et al. (1981a,b), as cited in ATSDR (1987), indicated that dermal exposure to PCBs at 2-28 $\mu g/cm^2$ of skin (on the hands) was not associated with clear evidence of liver disease, but may have been associated with liver enzyme induction in some of the workers. Assuming a total surface area for the hands of 910 sq. cm (Hawley 1985, as cited in ATSDR 1987) and body weight of 70 kg, the dermal exposure would have been 0.026-0.364 mg/kg/day. Because the workers were also exposed to PCBs by inhalation (48-275 $\mu g/m^3$), and because interpretation of the study is confounded by the lack of a control group, the dermal exposure range is not plotted on Figures 3.3 and 3.6.

Dermal application of Aroclor 1260 to rabbits on 5 day/wk at a dose of 118 mg/day for 38 days (27 total applications) produced degenerative lesions of the liver and kidneys, increased fecal porphyrin elimination, and hyperplasia and hyperkeratosis of the follicular and epidermal epithelium (Vos and Beems 1971, as cited in ATSDR 1987). As body weight appeared to be approximately 2.7 kg, the FEL of 118 mg/day is equal to a dose of 43.7 mg/kg/day (ATSDR 1987).

Developmental and Reproductive Toxicity (Dermal) - Pertinent data regarding developmental and reproductive effects of dermal exposure to PCBs were not located in the available literature (ATSDR 1987).

Genotoxicity (Dermal) - The PCBs have produced generally negative results in in vivo and in vitro genotoxicity tests (ATSDR 1987).

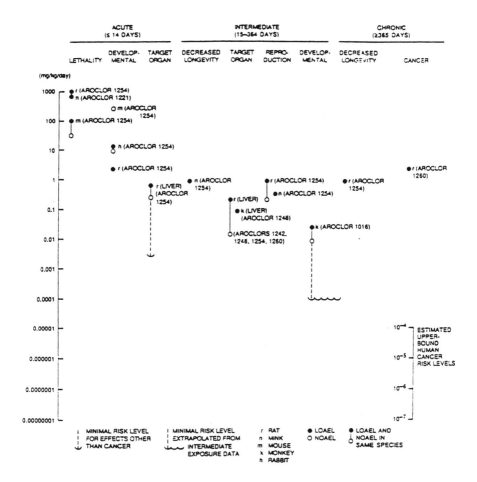

<u>FIGURE 3.5.</u> Levels of Significant Exposure for PCBs, Oral

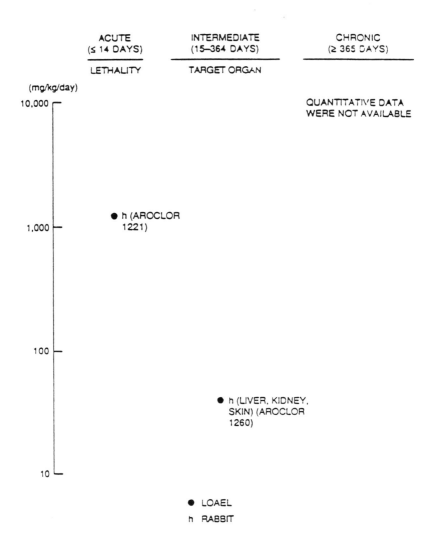

FIGURE 3.6. Levels of Significant Exposure for PCBs, Dermal

Carcinogenicity (Dermal) - Occupational exposure to PCBs, which involves inhalation as well as dermal exposure, provides inadequate evidence of carcinogenicity in humans. In two-stage carcinogenesis studies with mouse skin, Aroclor 1254 did not produce evidence of promoter or complete carcinogen activity and was not tested adequately for initiator activity (ATSDR 1987).

Human Carcinogenicity - Weight-of-Evidence Classification - The Weight-of-evidence classification for PCBs is "B2; probable human carcinogen." The basis is hepatocellular carcinomas in three strains of rats and two strains of mice and inadequate, yet suggestive, evidence of excess risk of liver cancer in humans by ingestion and inhalation or dermal contact (IRIS 1990).

The human carcinogenicity data are considered inadequate. Although many studies exist, the data are inadequate because of confounding exposures or lack of exposure quantification. The first documentation of carcinogenicity associated with PCB exposure was reported at a New Jersey petrochemical plant involving 31 research and development employees and 41 refinery workers (Bahn et al. 1976, 1977, as cited in IRIS 1990). Although a statistically significant increase in malignant melanomas was reported, the two studies failed to report a quantified exposure level and to account for the presence of other potential or known carcinogens. In an expanded report of these studies, NIOSH (1977), as cited in IRIS 1990, concurred with Bahn et al. findings (IRIS 1990).

Brown and Jones (1981), as cited in IRIS (1990), reported a retrospective cohort mortality study on 2567 workers who had completed at least 3 months of employment at one or two capacitor manufacturing plants. Exposure levels were 24-393 mg/m^3 at plant A and 318-1260 mg/m^3 at plant B. No excess risk of cancer was observed. In a 7-year follow-up study, Brown (1987), as cited in IRIS (1990), reported a statistically significant excess risk of liver and biliary cancer, with four of the five liver cancers in female workers at plant B. A review of the pathology reports indicated that two of the liver tumors counted in the follow-up study were not primary liver tumors. When these tumors are excluded, the elevation in incidence is not statistically significant. The results also may be confounded by population differences in alcohol consumption, dietary habits, and ethnic composition (IRIS 1990).

Bertazzi et al. (1987), as cited in IRIS (1990), conducted a mortality study of 544 male and 1556 female employees of a capacitor-making facility in Northern Italy. Aroclor 1254 and pyralene 1476 were used in this plant until 1964. These were progressively replaced by Pyralene 3010 and 3011 until 1970, after which lower chlorinated Pyralenes were used exclusively. In 1980 the use of PCBs was abandoned. Some employees also used TCE but, according to the authors, were "presumed" to be protected by efficient ventilation. Air samples were collected and analyzed for PCBs in 1954 and 1977 because of reports of chloracne in workers. Quantities of PCBs on workers' hands and workplace surfaces also were measured in 1977. In 18 samples, levels ranged from 0.2-159.0 $\mu g/m^2$ on workplace surfaces and 0.3 to 9.2 $\mu g/m^2$ on worker hands (IRIS 1990).

The authors compared observed mortality with that expected between 1946 and 1982 based on national and local Italian mortality rates. With vital status ascertainment 99.5% complete, relatively few deaths were reported by 1982 [30 males (5.5%) and 34 females (2.2%)]. In cohort males, the number of deaths from malignant tumors was significantly higher than expected compared with local or national rates, as was the number of deaths from cancer of the GI tract (6 observed vs. 1.7 national expected and 2.2 local expected). Of the six GI cancer deaths, one was caused by liver cancer and one by biliary tract cancer.

Deaths from hematologic neoplasms in males were also higher than expected, but excess was not statistically significant. Total cancer deaths in females were significantly elevated in comparison to local rates (12 observed vs. 5.3 expected). None of these were liver or biliary cancers. The number of deaths from hematologic neoplasms in females was higher than expected when compared with local rates (4 observed vs. 1.1 expected).

This study is limited by several factors, particularly the small number of deaths that occurred by the cut-off period. The power of the study is insufficient to detect an elevated risk of site-specific cancer. In addition, the authors stated that after an examination of the individual cases, interpretation of the increase in GI tract cancer in males was limited. This is because it appeared likely that some of these individuals had only limited

PCB exposure. Confounding factors may have included possible contamination of the PCBs by dibenzofurans and exposure of some of the workers to TCE, alkyl-benzene, and epoxy resins (IRIS 1990).

Two occurrences of ingestion of PCB-contaminated rice oil have been reported: the Yusho incident of 1968 in Japan and the Yu-Cheng incident of 1979 in Taiwan. Amano et al. (1984), as cited in IRIS (1990), completed a 16-year retrospective cohort mortality study of 581 male and 505 female victims of the Yusho incident. A consistently high risk of liver cancer in females over the entire 16 years was observed; liver cancer in males was also significantly increased.

Several serious limitations are evident in this study. There was a lack of information regarding job histories or the influence of alcoholism or smoking. The information concerning the diagnosis of liver cancer was obtained from the victims' families, and it is not clear whether this informa-tion was independently verified by health professionals. For some of the cancers described, the latency period is shorter than would be expected. Furthermore, the contaminated oils contained polychlorinated dibenzofurans and polychlorinated quinones as well as PCBs, and the study lacks data regarding exposure to the first two classes of compounds. There is strong evidence indicating that the health effects seen in Yusho victims were caused by ingestion of polychlorinated dibenzofurans, rather than to PCBs themselves (reviewed in EPA 1988, as cited in IRIS 1990). The results of the Amano et al. study can, therefore, be considered as no more than suggestive of carcinogenicity of PCBs (IRIS 1990).

Animal Carcinogenicity Data - The animal carcinogenicity data are con-sidered sufficient. PCB mixtures assayed in the following studies were commercial preparations and may not be the same as mixtures of isomers found in the environment. Although animal feeding studies demonstrate the carcino-genicity of commercial PCB preparations, it is not know which of the PCB congeners in such preparations were responsible for these effects, or if decomposition products, contaminants or metabolites were involved in the toxic response. Early bioassays with rats (Kimura and Baba, 1973; Ito et al., 1974, as cited in IRIS 1990) were inadequate to assess carcinogenicity because of the small number of animals and short duration of exposure to PCBs. A long-term bioassay of Aroclor 1260 reported by Kimbrough et al. (1975), as cited in IRIS (1990), produced hepatocellular carcinomas in female Sherman rats when 100 ppm was administered for 630 days to 200 animals. Hepatocellular carcinomas and neoplastic nodules were observed in 14 and 78%, respectively, of the dosed animals, compared with 0.58 and 0%, respectively, of the controls (IRIS 1990).

The National Cancer Institute (NCI) (1978), as cited in IRIS (1990), reported results for 24 male and 24 female Fischer 344 rats treated with Aroclor 1254 at 25, 50, or 100 ppm for 104 to 105 weeks. Although carcinomas of the GI tract were observed among the treated animals only, the incidence was not statistically significantly elevated. An apparent dose-related incidence of hepatic nodular hyperplasia in both sexes as well as hepato-cellular carcinomas among mid- to high-dose treated males was reported (4-12%, compared to 0% in controls).

Norback and Weltman (1985), as cited in IRIS (1990), fed 70 male and 70 female Sprague-Dawley rats a diet containing Aroclor 1260 in corn oil at 100 ppm for 16 months, followed by a 50-ppm diet for an additional 8 months, then a basal diet for 5 months. Control animals (63 rats/sex) received a diet containing corn oil for 18 months, then a basal diet alone for 5 months. Among animals that survived for at least 18 months, females exhibited a 91% incidence (43/47) of hepatocellular carcinoma. An additional 4% (2/47) had neoplastic nodules. In males corresponding incidences were 4% (2/46) for carcinoma and 11% (5/46) for neoplastic nodules. Concurrent liver morphology studies were carried out on tissue samples obtained by partial hepatectomies of three animals per group at eight time points. These studies showed the sequential progression of liver lesions to hepatocellular carcinomas (IRIS 1990).

Orally administered PCB resulted in increased incidences of hepato-cellular carcinomas in two mouse strains. Ito et al (1973), as cited in IRIS (1990), treated male dd mice (12/group) with Kanechlors 500, 400 and 300 each at dietary levels of 100, 250, or 500 ppm for 32 weeks. The group fed 500 ppm of Kanechlor 500 had a 41.7% incidence of hepatocellular carcinomas and a

58.3% incidence of nodular hyperplasia. Hepatocellular carcinomas and nodular hyperplasia were not observed in mice fed 100 or 250 ppm of Kanechlor 500, nor among those fed Kanechlors 400 or 300 at any concentration (IRIS 1990).

Schaeffer et al. (1984), as cited in IRIS (1990), fed male Wistar rats diets containing 100 ppm of the PCB mixtures Clophen A 30 (30% chlorine by weight) or Clophen A 60 (60% chlorine by weight) for 800 days. The PCB mixtures were reported to be free of furans. Clphen A 30 was administered to 152 rats, Clophen A 60 to 141 rats, and 139 rats received a standard diet. Mortality and histologic lesions were reported for animals necropsied during each 100-day interval for all three groups.

Of the animals that survived the 800-day treatment period, 1/53 rats (2%) in the control group, 3/87 (3%) in the Clophen A 30 group and 52/85 (61%) in the Clophen A 60 group were significantly elevated in comparison to the control group. Neoplastic liver nodules were reported in 2/53 control, 35/87 Clophen A 30, and 34/85 Clophen A 60-treated animals. The incidence of nodules was significantly increased in both treatment groups in comparison to the control group. Neoplastic liver nodules and hepatocellular carcinomas appeared earlier and at higher incidence in the Clophen A 60 group relative to the Clphen A 30 group. The authors interpreted the results as indicative of a carcinogenic effect related to the degree of chlorination of the PCB mixture. The authors also suggested that these findings support those of others, including Ito et al. (1973) and Kimbrough et al. (1975), both as cited in IRIS (1990), in which hepatocellular carcinomas were produced by more highly chlorinated mixtures (IRIS 1990).

Kimbrough and Linder (1974), as cited in IRIS (1990), dosed groups of 50 male BALB/cJ mice (a strain with a low spontaneous incidence of hepatoma) with Aroclor 1254 at 300 ppm in the diet for 11 months or 6 months, followed by a 5-month recovery period. Two groups of 50 mice were fed a control diet for 11 months. The incidence of hepatomas in survivors fed Aroclor 1254 for 11 months was 10/22. One hepatoma was observed in the 24 survivors fed Aroclor 1254 for 6 months (IRIS 1990).

Supporting Data for Carcinogenicity - Most genotoxicity assays of PCBs have been negative. The majority of microbial assays of PCB mixtures and various congeners showed no evidence of mutagenic effects (Schoeny et al., 1979; Schoeny, 1982; Wyndham et al., 1976, all as cited in IRIS 1990). Of various tests on the clastogenic effect of PCBs (Heddle and Bruce, 1977; Green et al., 1975, as cited in IRIS 1990), only Peakall et al. (1972), as cited in IRIS (1990), reported results indicative of a possible clastogenic action by PCBs in dove embryos (IRIS 1990).

Chlorinated dibenzofurans (CDFs), known contaminants of PCBs, and chlorinated dibenzodioxins (CDDs) are structurally related to, and produce certain biologic effects similar to, those of PCB congeners. While CDDs are known to be carcinogenic, the carcinogenicity of CDFs is still under evaluation (IRIS 1990).

Quantitative Estimate of Carcinogenic Risk - The Oral Slope Factor for PCBs is 7.7/mg/kg/day. The Drinking Water Unit Risk is 2.2E-4/μg/L. The extrapolation method used was the linear multistage procedure with extra risk. Drinking water concentrations at specified risk levels is as follows:

Risk Level	Concentration
E-4 (1 in 10,000)	5E-1 μg/L
E-5 (1 in 100,000)	5E-2 μg/L
E-6 (1 in 1,000,000)	5E-3 μg/L

Dose-Response Data (Carcinogenicity, Oral Exposure) - The tumor type developed was trabecular carcinoma/adenocarcinoma, neoplastic nodule. The test animals were female Sprague-Dawley rats. The route of exposure was an oral diet. The dose-response data are provided in Table 3.1 (Norback and Weltman, 1985, as cited in IRIS 1990).

Human equivalent dosage assumes a TWA daily dose of 3.45 mg/kg/day. This reflects the dosing schedule of 5 mg/kg/day (assuming the rat consumes an amount equal to 5% of its body weight per day) for the first 16 months, 2.5 mg.kg/day for the next 8 months, and no dose for the last 5 months (IRIS 1990).

TABLE 3.1. PCB Oral Exposure Carcinogenicity Dose-Response Data

Administered (mg/kg/day) (TWA)	Human Equivalent (mg/kg/day)	Tumor Incidence
0	0	1/49
3.45	0.59	45/47

A slope factor of 3.9/mg/kg/day was based on data from a study by Kimbrough et al. (1975), as cited in IRIS (1990), of female Sherman rats fed Aroclor 1260. The estimate based on the data of Norback and Weltman (1985), as cited in IRIS (1990), is preferred because Sprague-Dawley rats are known to have low incidence of spontaneous hepatocellular neoplasms. Moveover, the latter study spanned the natural life of the animal, and concurrent morphologic liver studies showed the sequential progression of liver lesions to hepatocellular carcinomas (IRIS 1990).

Though it is known that PCB congeners vary greatly as to their potency in producing biological effects, for purposes of this carcinogenicity assessment, Aroclor 1260 is intended to be representative of all PCB mixtures. There is some evidence that mixtures containing more highly chlorinated biphenyls are more potent inducers of hepatocellular carcinoma in rats than mixtures containing less chlorine by weight (reviewed in Kimbrough 1987 and Schaefer et al. 1984, as cited in IRIS 1990).

The unit risk should not be used if the water concentration exceeds 50 μg/L, because above this concentration, the slope factor may differ from that stated (IRIS 1990).

No quantitative estimate of carcinogenic risk from inhalation exposure of PCBs is available at this time.

4. Chlorofluorocarbons

Chlorofluorocarbons (CFCs) are in common use in public buildings as propellents and refrigerants. These CFCs have the potential to impact human health in two ways: 1) the toxicological effects of the contaminants themselves, and 2) the health impact associated with their role in reducing the stratospheric ozone layer. Considerable research and controversy are occurring on the reduction of the stratospheric ozone layer. In this report, three CFCs were selected to examine direct health effects: 1,1,2-trichloro-1,2,2-trifluoroethane (CFC-113), trichlorofluoromethane, and dichlorodifluoromethane. Because the human health impact of the reduction of the stratospheric ozone layer is part of a very involved and ongoing investigative process, only a very general discussion of the potential health effects from the reduction of the stratospheric ozone layer will be provided.

4.1 DIRECT HEALTH EFFECTS OF CFCs

The EPA-maintained IRIS system addresses three CFCs of concern: 1,1,2-trichloro-1,2,2-trifluoroethane (CFC-113 or FREON 113), trichlorofluoromethane (Freon 11), and dichlorodifluoromethane (FREON 12).

4.1.1 1,1,2-Trichloro-1,2,2-Trifluoroethane (CFC-113 or FREON 113)

Freon 113 has a Hazard Rating of 1 (Sax and Lewis 1989).

Noncarcinogenic Effects - The RfD is based on the assumption that thresholds exist for certain toxic effects, such as cellular necrosis, but may not exist for other toxic effects such as carcinogenicity. In general, the RfD is an estimate of a daily exposure to the human population (including sensitive subgroups) that is likely to be without an appreciable risk of deleterious effects during a lifetime. The RfDo for trichlorotrifluoroethane is 3E+1 mg/kg/day, with the critical effect being psychomotor impairment. The NOAEL is 5358 mg/m^3 (can be converted to 273 mg/kg/day) (IRIS 1990). IRIS noted that the RfDo for trichlorotrifluoroethane may change in the near future, pending the outcome of a further review being conducted by the Oral RfD Work Group.

Several animal inhalation studies reported negative results in dogs, rabbits, and rats chronically exposed to very high concentrations of trichlorotrifluoroethane (EPA 1983, as cited in IRIS 1990). No apparent adverse effects have been reported in humans occupationally exposed to trichlorotrifluoroethane at either 500 mg/m^3 levels for 11 years of 5358 mg/m^3 for 2.77 years (Imbus and Adkins, 1972, as cited in IRIS 1990).

Slight impairment of psychomotor performance was reported in male volunteers exposed to trichlorotrifluoroethane concentrations of 19,161 mg/m^3 for 2.75 hours (Stopps and McLaughlin 1967, as cited in IRIS 1990). This exposure period was too brief to consider a NOAEL for chronic exposure. Therefore, the RfD of 30 mg/kg/day is considered protective (IRIS 1990).

There is no RfDi for trichlorotrifluoroethane at this time.

The confidence in the RfDo is low. Confidence in the chosen study, database, and RfDo are all considered low. Despite the fact that the chosen study describes human data and the fact that several chronic studies in animals are supportive, uncertainties in both the exposure levels and route extrapolation preclude higher confidence ratings (IRIS 1990).

Human Carcinogenicity - Trichlorotrifluoroethane has not been evaluated by the EPA for evidence of human carcinogenic potential (IRIS 1990).

4.1.2 Trichlorofluoromethane (FREON 11)

FREON 11 has a Hazard Rating of 3 (Sax and Lewis 1989).

Noncarcinogenic Effects - The RfDo for trichlorofluoromethane is 3E-1 mg/kg/day, with survival and histopathology as the critical effects. No NOAEL was established. The LOAEL is 488 mg/kg/day (can be converted to 349 mg/kg/day) (IRIS 1990).

The NCI bioassay (NCI 1978, as cited in IRIS 1990) was performed on rats and mice exposed to various doses of trichloromonofluoromethane by gavage over a period of 78 weeks (50 animals/species/sex/dose for each of two doses with 20 animals/species/sex for each of two control groups). A statistically significant positive association between increased dosage and accelerated mortality by the Tarone test in male and female rats and female mice was observed. In treated rats of both sexes, there were also elevated incidences

of pleuritis and pericarditis not seen in controls. Inhalation studies that
employed multispecies exposures to higher levels of the compound than used by
the NCI (Leuschner et al. 1983, Colman et al. 1981, Hansen et al. 1984, all as
cited in IRIS 1990) reported no adverse clinical/pathologic signs of toxicity
from subchronic or short-term exposures (IRIS 1990).

The LOAEL of 488 mg/kg/day (mortality in rats) was converted to 349 mg/
kg/day on a 7-day exposure basis (IRIS 1990).

The RfDo has been given a confidence level of medium. The chosen study
is given a medium confidence rating because large numbers of animals per sex
were tested in two doses for chronic exposures, but the study did not
establish a NOAEL. The database is given a medium confidence rating because
of the support of chronic data but lack of reproductive data. A medium
confidence in the RfDo follows (IRIS 1990).

No RfDi is available at this time.

Human Carcinogenicity - Trichlorofluoromethane has not been evaluated by
the EPA for evidence of human carcinogenic potential (IRIS 1990).

4.1.3 Dichlorodifluoromethane (FREON F-12)

FREON F-12 has a Hazard Rating of 1 (Sax and Lewis 1989).

Noncarcinogenic Effects - The RfDo for dichlorodifluoromethane is 2E-1
mg/kg/day, with reduced body weight the critical effect. The NOEL is 300 ppm
(can be converted to 15 mg/kg/day). The LOAEL is 3000 ppm (can be converted
to 150 mg/kg/day) (IRIS 1990).

The study reported by the Haskell Laboratory (Sherman 1974, as cited in
IRIS 1990) involved 2-year feeding studies in which dogs and rats received 300
ppm or 3000 ppm of dichlorodifluoromethane. This report contained data on
clinical biochemical, urine analytical, hematological or histopathological
evaluations. Additionally, carcinogenic and three-generation reproductive
studies were conducted in rats. Except for decreased weight gain in rats
(about 20% in females) that received 3000 ppm dichlorodifluoromethane in the
diet, no other adverse effects were attributable to this compound in either
rats or dogs (IRIS 1990).

The Haskell Laboratory study is sufficiently complete to derive an RfD for adequate protection against adverse human health effects. The high dose (3000 ppm) caused decreased body weights in rats and is therefore considered a LOAEL, whereas the low dose (300 ppm) in rats produced no adverse effects attributable to the oral administration of dichlorodifluoromethane (IRIS 1990).

The confidence in the RfDo is medium. The Haskell Laboratory study is a chronic oral study in two species that incorporated extensive clinical and toxicologic parameters. Therefore, a high level of confidence in the study is appropriate. Confidence in the database is medium because of the lack of teratology and reproductive data. Medium confidence in the RfDo follows (IRIS 1990).

There is no RfDi available at this time.

4.2 HEALTH EFFECTS OF STRATOSPHERIC OZONE LAYER REDUCTION

The stratospheric ozone layer has several interesting effects. Because of the absorption characteristics of ozone, it filters ultraviolet radiation on its way to the earth's surface. Thus, it reduces the amount of photo-chemical activity to which humans are exposed. This reduction is essential because humans can not survive exposure to the ultraviolet radiation from the sun without some attenuation by the ozone layer. Consequently, any signifi-cant reduction of this ozone layer could directly affect human health.

Nevertheless, humans need some ultraviolet radiation. The amount of sunlight humans are exposed to directly affects the amount of vitamin D produced in the skin. A substantial reduction in ultraviolet radiation would lead to vitamin D deficiencies in humans. On the other hand, if the ultra-violet radiation increased considerably, severe skin problems would result. The vast problem of global warming and all of its impacts on humans are also part of the concern over the depletion of the ozone layer.

The CFCs considered are only one of several contributors to the problem of the stratospheric ozone layer depletion. For example, proposed supersonic jet flights in the upper atmosphere have been a major concern. Nitric oxides released from these planes react with ozone to reduce the average ozone concentration at those altitudes.

5. Implications of Risk Assessment in Indoor Air

The number of hazardous chemical contaminants potentially present in the indoor air of public buildings is overwhelming. However, the overall health impact is probably not as great as would be expected because of the normally low concentrations of these chemicals in the indoor air. The information needed to adequately assess the magnitude of the health impact of each of these chemicals is quite limited. In addition, these chemicals exist in the indoor air environment in the form of complex mixtures, which in many cases causes the chemicals to behave differently (in terms of effect on human health) than they do when considered separately.

The health effects information presented in this document reflects current research and conclusions, but several limitations exist. First, health effect data was normally determined by exposing animals to fairly high doses of chemicals. This approach requires that the data be extrapolated back down to the normally low doses of the chemical that a person is exposed to in public buildings. This extrapolation process is complicated by the assumptions that must be made; for example, whether there is a threshold effect for the chemical and whether the data is linear at low doses or how it should be extrapolated for these low doses.

This data interpretation process is further complicated by the fact that the data must be translated from the effects on animals to the effects on humans. In addition, as mentioned, chemicals exist in the indoor air environment in the form of complex mixtures. Considerably more research needs to be conducted to draw definitive conclusions as to the full impact on human health by the host of potential hazardous chemicals that may be found in the indoor environment of public buildings.

Normally, the assumptions made regarding the development of the health effect parameters and factors presented in this document are conservative (i.e., they err on the side of overpredicting adverse effects). Thus, while research continues on health effects of the host of chemicals and complex chemical mixtures that could exist in public buildings, these parameters and factors (which should be updated regularly as research progresses) should be

285

used to provide estimates of the health impact and risk for particular building situations. In making building-specific health or risk assessments, specific building conditions should be determined as accurately as possible to consider the most realistic set of chemicals, mixture combinations, concentrations of chemicals, and physical conditions.

6. References

Agency for Toxic Substances and Disease Registry (ATSDR). 1987. *Toxicological Profile for Selected PCBs*. U.S. Public Health Service, Centers for Disease Control, Agency for Toxic Substances and Disease Registry, Atlanta, Georgia.

American Conference of Governmental Industrial Hygienists (ACGIH). 1988. *Threshold Limit Values and Biological Exposure Indices for 1988-1989*. ISBN: 0-936712-78-3. Cincinnati, Ohio.

Baechler, M. C., H. Ozkaynak, J. D. Spengler, L. A. Wallace, and W. C. Nelson. 1989. "Assessing Indoor Exposure to Volatile Organic Compounds Released From Paint Using the NASA Data Base." In *Proceedings of 82nd Annual Meeting and Exhibition, Air and Waste Management Association*, Anaheim, California.

Girman, J. R., and A. T. Hodgson. 1986. "Source Characterization and Personal Exposure to Methylene Chloride from Consumer Products." In *Proceedings of the 79th Annual Meeting of Air Pollution Control Association (APCA)*, Minneapolis, Minnesota.

Girman, J. R., A. T. Hodgson, A. S. Newton, and A. W. Winkes. 1987. "Volatile Organic Emissions from Adhesives with Indoor Applications." In *Proceedings of the International Conference on Indoor Air Quality and Climate, Institute for Water, Soil and Air Hygiene*, Berlin.

Hazardous Substances Data Bank (HSDB). 1990. *On Line (An Active On-Line Data Base Available on the MEDLARS System)*. Maintained by National Library of Medicine). National Library of Medicine, MEDLARS Management Section, Rockville, Maryland.

Integrated Risk Information System (IRIS). 1990. *On Line (An Active On-Line Data Base Available on the TOXNET System)*. Maintained by EPA. U.S. Environmental Protection Agency, Office of Health and Environmental Assessment, Environmental Criteria and Assessment Office, Cincinnati, Ohio.

Jungers, R. H., and L. S. Sheldon. 1987. "Characterization of Volatile Organic Chemicals in Public Access Buildings." In *Indoor Air '87, Proceedings of the 4th International Conference on Indoor Air Quality and Climate, Institute for Water, Soil and Air Hygiene*, Berlin.

Kalf, G. F., G. B. Post, and R. Snyder. 1987. "Solvent Toxicology: Recent Advances in the Toxicology of Benzene, The Glycol Ethers, and Carbon Tetrachloride." *Ann. Rev. Pharmacol. Toxicol.* pp. 399-427, Annual Review Incorporated.

Krause, C., W. Mailahn, R. Nagel, C. Schulz, B. Seifert, and D. Ullrreh. 1987. "Occurrence of Volatile Organic Compounds in the Air of 500 Homes in the Federal Republic of Germany." In *Indoor Air '87, Proceedings of the 4th International Conference on Indoor Air Quality and Climate, Institute for Water, Soil and Air Hygiene*, Berlin.

Krzyzanowski, M., J. J. Quackenboss and M. D. Lebowitz. 1989. *Chronic Respiratory Effects of Indoor Formaldehyde Exposure*. University of Arizona Health Sciences Center, Tucson, Arizona.

Molhave, L. 1982. "Indoor Air Pollution Due to Organic Gases and Vapours of Solvents in Building Materials." *Environment International*, 8:117-127.

National Institutes of Health (NIH). 1989. *National Library of Medicine: MEDLARS The World of Medicine At Your Fingertips*. NIH Publication No. 89-1286, U.S. Department of Health and Human Services, Bethesda, Maryland.

Nied, G. J., and P. K. TerKonda. 1983. *Health Effects from Formaldehyde Emissions*. Department of Health, Division of Air Pollution Control, Cleveland, Ohio and University of Missouri-Rolla, Rolla, Missouri.

Registry of Toxic Effects of Chemical Substances (RTECS). 1987. *Hard-Copy (An Active Online Data Base Available through National Institute of Occupational Safety and Health [NIOSH])*, Cincinnati, Ohio.

Rinsky, R. A., R. J. Young, and A. B. Smith. 1981. "Leukemia in Benzene Workers." *Am. J. Ind. Med.* 2:217-245.

Rinsky, R. A., A. B. Smith, R. Hornung, T. G. Filloon, R. J. Young, A. H. Okun, and P. J. Landrigan. 1987. "Benzene and Leukemia: An Epidemiologic Risk Assessment." *N. Engl. J. Med.* 316(17):1044-1050.

Sax, N. I. and R. J. Lewis, Sr. 1989. *Dangerous Properties of Industrial Materials*, Volumes I, II, and III. Van Nostrand Reinhold, New York, New York.

Tichenor, B. A., and M. A. Mason. 1986. "Characterization of Organic Emissions from Selected Materials in Indoor Use." In *Proceedings of the Air Pollution Control Association (APCA), 79th Annual Meeting*, Minneapolis, Minnesota.

Tichenor, B. A., and M. A. Mason. 1988. "Organic Emissions from Consumer Products and Building Materials to the Indoor Environment." *Journal of the Air Pollution Control Association (APCA)*, 38: 264-268.

Tichenor, B. A. 1989. *Indoor Air Sources: Using Environmental Test Chambers to Characterize Organic Emissions from Indoor Materials and Products*. EPA-600/8-89-074, U.S. Environmental Protection Agency, Air and Energy Engineering Research Laboratory, Research Triangle Park, North Carolina.

U.S. Department of Energy (DOE). 1984. *The Expanded Residential Weatheriza-tion Program - Final Environmental Impact Statement*. DOE/EIS-0095F, Bonneville Power Administration, Portland, Oregon.

U.S. Department of Energy (DOE). 1987. *Draft Environmental Impact Statement on New Energy-Efficient Homes Programs, Volume II*. DOE/EIS-0127, Bonneville Power Administration, Portland, Oregon.

U.S. Environmental Protection Agency (EPA). 1984a. *Health Effects Assessment for Benzene*. EPA/540/1-86/037, U.S. EPA, Office of Research and Development, Cincinnati, Ohio, and Office of Emergency and Remedial Response, Washington, D.C.

U.S. Environmental Protection Agency (EPA). 1984b. *Health Effects Assessment for Carbon Tetrachloride*. EPA/540/1-86/039, U.S. EPA, Office of Research and Development, Cincinnati, Ohio.

U.S. Environmental Protection Agency (EPA). 1984c. *Health Effects Assessment for Chloroform*. EPA/540/1-86/010, U.S. EPA, Office of Research and Develop-ment, Cincinnati, Ohio.

U.S. Environmental Protection Agency (EPA). 1985. *Health Assessment Document for Tetrachloroethylene (Perchloroethylene)* - Final Report. EPA/600/s-82/005F, U.S. EPA, Office of Research and Development, Research Triangle Park, North Carolina.

U.S. Environmental Protection Agency (EPA). 1987. *Health Effects Assessment for Trimethylbenzenes*. EPA/600/8-88/060, U.S. EPA, Office of Research and Development, Cincinnati, Ohio.

U.S. Environmental Protection Agency (EPA). 1988. *Indoor Air Facts - Sick Buildings*. U.S. EPA, Office of Research and Development and Office of Air and Radiation, Washington, D.C.

U.S. Environmental Protection Agency (EPA). 1988. Appendix to EPA 1985.

U.S. Environmental Protection Agency (EPA). 1988b. *Indoor Air Quality in Public Buildings: Volume I*. EPA/600/S6-88/009, U.S. EPA, Office of Acid Deposition, Environmental Monitoring and Quality Assurance, Washington, D.C.

U.S. Nuclear Regulatory Commission (NRC). 1986. *Environmental Tobacco Smoke: Measuring Exposures and Assessing Health Effects*. National Research Council, National Academy Press, Washington, D.C.

Wallace, L. A., E. D. Pellizzari, and S. M. Gordon. 1984. "Organic Chemicals in Indoor Air: A Review of Human Exposure Studies and Indoor Air Quality Studies" in *Indoor Air and Human Health. Seventh Life Sciences Symposium*. Oak Ridge National Laboratory, Lewis Publishers, Inc., Chelsea, Michigan.

Wallace, L. A. 1986. "Estimating Risk from Measured Exposures to Six Suspected Carcinogens in Personal Air and Drinking Water in 600 U.S. Residents." In *Proceedings of Air Pollution Control Association (APCA), 79th Annual Meeting*, Minneapolis, Minnesota.

Wallace, L. A. 1987. *The Total Exposure Assessment Methodology (TEAM) Study Summary and Analysis: Volume 1*. EPA/600/6-87/002, U.S. EPA, Office of Research and Development, Washington, D.C.

Wallace, L. A., R. Jungers, L. Sheldon, and E. Pellizzari. 1987a. "Volatile Organic Chemicals in 10 Public-Access Buildings." In *Indoor Air '87: Proceedings of the 4th International Conference on Indoor Air Quality and Climate*. Institute for Water, Soil and Air Hygiene, Berlin.

Wallace, L. A., E. Pellizzari, B. Leaderer, H. Zelon, and L. Sheldon. 1987b. "Emissions of Volatile Organic Compounds from Building Materials and Consumer Products." *Atmospheric Environment 21:385-393, Pergamon Journals Ltd.*

Walsh, P. J., C. S. Dudney, and E. D. Copenhaver. 1984. *Indoor Air Quality*. CRC Press, Inc., Boca Raton, Florida.

Part III

Suggested Methods of Analysis for Indoor Air Environmental Carcinogens

The information in Part III is from *Indoor Air—Assessment: Methods of Analysis for Environmental Carcinogens,* prepared by Max R. Peterson and Dennis F. Naugle of Research Triangle Institute and Michael A. Berry of the U.S. Environmental Protection Agency Environmental Assessment and Criteria Office, for the U.S. Environmental Protection Agency, June 1990.

DISCLAIMER

Introduction

Indoor air compounds with carcinogenic activity in animals, humans, or both include radon, asbestos, organic compounds, inorganic compounds, particles (including environmental tobacco smoke or ETS), and non-ionizing radiation (NIR). Some of the pollutants in this list exist as vapors, some as fibers or particulate material, some are adsorbed onto suspended particulate material, and some are distributed between a vapor-phase and a particle-bound state. Some classes contain a large number of compounds (ETS and organics), one is associated with a single parent element (radon); and one is not a chemical species at all (NIR). Most are dangerous in themselves but one (radon) is a precursor to even more deadly progeny. Because of these differences in physical and chemical states and properties, each carcinogen class generally requires different sampling and analysis approaches.

As a further complication, a single indoor environment may contain a wide variety of air pollutant mixtures. No single best approach currently exists for assessing the composition and concentrations of such complex mixtures. There are many uncertainties inherent in characterizing the pollutant mix of indoor environments and there is great difficulty in predicting the severity and nature of toxicological interactions.

Identical mixtures of pollutants from different sampling sites are rarely, if ever, encountered in indoor air. Similar mixtures may be found in a variety indoor environments, but with concentrations of individual components differing significantly. For example, ETS may contain 4,500 individual compounds, each present at a concentration that depends upon a wide variety of factors. Therefore, from a practical sampling and analysis point of view, only the most toxic of the compounds or the pollutants making up the largest portions of the mixture can be measured.

The analytical methods presented below are classified by pollutant category with sub-categories added as deemed appropriate. The purpose of this monograph is to present an overview of some common methods for each category of pollutant or, in some cases, for specific pollutants. While the list of methods for each category is not exhaustive, significant effort was expended to include at least one or two workable methods for each category.

292

Radon

OVERVIEW

Sampling and analysis of radon and radon progeny (decay products) in indoor air are complicated by both spatial and temporal changes in concentration. An acceptable measurement scheme must also take into account diurnal and seasonal variations.

In addition, analysis is complicated by dramatic changes in composition of a collected sample with time. The change in composition is due to the radioactive decay of both radon and its progeny. Radon decays more slowly (half-life 3.83 days) than any of its progeny but still at a fast enough rate to require that analyses be made very soon after samples are taken (Cothern and Smith, 1987). New methodologies to measure time-integrated radon concentrations on indoor glass surfaces (such as picture glass) are under development and appear promising for long term human exposure studies (Samuelsson, 1988).

SAMPLING

Collection of samples may be accomplished by grab sampling, continuous sampling, or integrative sampling. Grab sampling involves the taking of essentially an instantaneous sample and allows determination of the concentration at the specific time the sample was taken. Continuous sampling involves the taking of many measurements at closely spaced time intervals and allows the determination of patterns in the variation of concentration over the entire sampling time. Integrative sampling involves the taking of a single sample over a long time interval and allows determination of a single average concentration over the sampling period (Cothern and Smith, 1987).

Radon may be separated from its progeny during sampling, and the concentration of either or both determined by analysis. The separation is typically accomplished by allowing radon to diffuse through a passive barrier (e.g., foam rubber). Radon progeny, which rapidly adsorb onto the surfaces of airborne particulates and other solids, cannot pass through such barriers.

293

ANALYSIS

Most of the analytical methods for radon and radon progeny actually measure emitted radiation and not concentration of the target species, although the two are directly related in the absence of other radioactive material. The decay chain for radon and its progeny is summarized in Table 1. The decay sequence includes the emission of three fundamental kinds of radioactivity: Alpha particles, beta particles, and gamma rays. The emission rate of any or all of these may be measured with appropriate instrumentation (Cothern and Smith, 1987).

Concentrations of radon are usually expressed in bequerels per cubic meter (Bq/m^3) or in picocuries per liter (pCi/L). One bequerel equals 1 event per second (decay/s) and one picocurie equals 0.037 bequerel.

Radon progeny concentrations are usually expressed in working level months (WLM). One working level equals the quantity of short-lived progeny that will result in 1.3×10^5 MeV of potential alpha energy per liter of air. Potential alpha energy is used to relate atmospheric concentration of radon progeny to dose delivered to the human lung. It represents the total alpha energy an atom can emit as it decays through its entire radioactive series, and it must be calculated from measurements (or estimates) of the concentrations of the progeny.

Although a large number of devices are available, the actual analysis for radon and/or its progeny is typically accomplished using a scintillation phosphor mounted on a photomultiplier, an ionization chamber, a thermoluminescent dosimeter (TLD), or a visual or automatic image-processing method. Table 2 provides a summary of currently available radon measurement devices based on instrument type.

Scintillation devices include cells for gaseous and liquid samples and plates for samples of radon progeny collected on filters. Gas scintillation cells and scintillation plates are typically coated with zinc sulfide; in liquid scintillation cells, the liquid contains both the radioactive material and the scintillator. In all cases, the intensity of the light pulses generated by the impingement of radioactive particles (alpha, beta, and/or gamma) on the scintillating material is measured with the aid of a photomultiplier tube (Cothern and Smith, 1987; Henschel, 1988).

An ionization chamber may be used to measure radon in air. The current in the chamber is directly proportional to the radon concentration (Cothern and Smith, 1987).

TABLE 1. RADON-222 DECAY SERIES

Isotope	Symbol	Decay Mode	Half-Life
Radon-222	$^{222}_{86}Rn$		2.82 days
	$\downarrow$	Alpha (^4_2He)	
Polonium-218*	$^{218}_{84}Po$		30.5 min
	$\downarrow$	Alpha (^4_2He)	
Lead-214*	$^{214}_{82}Pb$		26.8 min
	$\downarrow$	Beta ($^0_{-1}e$)	
Bismuth-214*	$^{214}_{83}Bi$		19.7 min
	$\downarrow$	Beta ($^0_{-1}e$)	
Polonium-214*	$^{214}_{84}Po$		0.000164 sec
	$\downarrow$	Alpha (^4_2He)	
Lead-210	$^{210}_{82}Pb$		21 years
	$\downarrow$	Beta ($^0_{-1}e$)	
Bismuth-210	$^{210}_{83}Bi$		5 days
	$\downarrow$	Beta ($^0_{-1}e$)	
Polonium-210	$^{210}_{84}Po$		138 days
	$\downarrow$	Alpha (^4_2He)	
Lead-206	$^{206}_{82}Pb$		Stable

*Radon progeny of primary concern.

A thermoluminescent dosimeter (TLD) may be used to provide an integrated measurement of radon or radon progeny activity. The TLD chip is typically lithium fluoride or calcium fluoride (Cothern and Smith, 1987).

In the two-filter method, a small tube, 30-100 cm long, is equipped with two filters: An entrance filter, which removes progeny from the sample as it is drawn into the tube, and an exit filter, which traps the polonium-218 formed by the decay of some of the radon atoms

TABLE 2. APPLICATIONS AND SENSITIVITIES OF SOME RADON
MEASUREMENT DEVICES

Instrument type	Application	Sensitivity[a]	Purpose[b]
Direct measurement			
Scintillation cell	Grab or continuous	3.7 Bq/m^{3c}	Screening, diagnostic
Ionization chamber	Grab or continuous	3.7 Bq/m^{3c}	Screening, diagnostic
Passive barrier method[d]			
Scintillator	Continuous	3.7 Bq/m^{3c}	Screening, diagnostic
TLD chip	Integrating	$0.08\text{-}8.1 \text{ Bq/m}^3$	Screening, large scale survey
Two-filter method	Grab or continuous	3.7 Bq/m^{3c}	Diagnostic
Passive sampling devices			
Activated charcoal	Integrating	7.4 Bq/m^3 for 100-hour exposure	Screening, large scale survey
Alpha track	Integrating	18.5 Bq/m^3 for 30-day exposure[c]	Screening, large scale survey

[a]With collection of progeny on (or in close proximity to) a scintillator or thermoluminescent detector (TLD) chip.
[b]Adapted from Cothern & Smith (1987). 1 pCi/L = 37 Bq/m^3.
[c]A sensitivity less than the value shown is generally achievable depending on the specific instrument used.
[d]Three typical purposes are illustrated:

- Screening: The aim of a screening method is to evaluate rapidly and inexpensively where high radon concentrations may occur.
- Diagnostic: Designed to measure specific parameters for detailed radon analysis such as:
  - short-term spatial and temporal variations,
  - relationship with other factors (such as ventilation rate),
  - equilibrium fraction of each of the radon daughters,
  - effect of remedial actions.
- Large-scale survey: A national, regional or other large study aimed at evaluating the exposure of the public. A large number of time-averaged measurements are needed.

as they move through the tube. After sampling, the filter is removed and counted by measurement of the alpha decays. The method has been adapted to continuous and integrative monitoring through the use of a TLD chip at the exit filter (Cothern and Smith, 1987).

Radon may be collected from air by adsorption onto the surface of activated charcoal. It may be analyzed by de-emanation of the radon from the charcoal into a scintillation cell and alpha-counting; by heating the charcoal and counting gamma emissions from the desorbed radon using a sodium iodide system; or by dissolving the charcoal in liquid scintillation fluid and counting in a liquid scintillation detector (Cothern and Smith, 1987).

In an alpha track detector, a thin piece of an appropriate plastic is placed in a holder and exposed to air containing radon (and perhaps its progeny) for an extended period of time (up to a year). During the exposure period, alpha particles, emitted by decaying nuclei, strike the surface of the plastic making microscopic gouge marks or alpha tracks. The exposed plastic is removed from its holder and chemically etched to enlarge the tracks. The alpha tracks, now appearing as small holes on the surface, are then visually counted using a wide-screen microscope (Cothern and Smith, 1987) or some other counting system (e.g., spark counter, CCD camera, etc.).

The principles used for the measurement of radon progeny are similar to those used to measure radon, but the sampling and analysis approach must be altered significantly. Radon progeny may be removed by passing the air sample through a filter. Radon gas passes through such a barrier while the progeny, which are adsorbed onto suspended particulate material, are collected on the filter.

The early methods for measuring radon progeny were based on the method originally reported by Kusnetz (1956) and Tsviglou et al. (1953). The progeny, which are largely attached to aerosols, are collected on a filter and subsequently analyzed by alpha-particle spectroscopy. This allows the determination of the activity of each progeny on the filter and the subsequent computation of the potential alpha energy or working level concentration.

Both passive and active alpha track detection have been applied to the measurement of radon progeny. Unfortunately, while progeny are easily excluded from the analysis for radon by this method, radon is not easily excluded from the analysis for progeny. The typical approach is to measure radon-plus-progeny, then radon only, and subtract to get the progeny-only value.

Programs designed to assist commercial firms offering radon measurement services have been established in many countries. In the United States, the Environmental Protection Agency (U.S. Environmental Protection Agency) supports a National Radon Measurement Proficiency (RMP) Program (U.S. Environmental Protection Agency, 1988). Each company participating in the National RMP Program submits passive detectors for exposure in a chamber containing a known concentration of radon or, in the case of continuous or active monitors, carries the device(s) to the chamber and samples the contaminated air directly through a port. After exposure, the passive detectors are returned to the company for analysis. The company reports its measured values for all detectors and/or monitors to the RMP Program Coordinator. The RMP Program currently uses the mean of the absolute value of the relative error (MARE) to evaluate the company's performance with a particular device or type of detector. The company is judged proficient at measurement if its MARE is 0.25 or less. The results of the RMP Program provide useful information on the relative accuracy of methods used to sample and measure radon and radon progeny in indoor air. Table 3 summarizes the results of the most recent proficiency measurement round for which published data are available (Singletary, 1988).

TABLE 3. RESULTS FROM ROUND 5 OF THE RADON
MEASUREMENT PROFICIENCY PROGRAM[a]

Species Method	Number[b] of sets	Range of target values	Median error[c] (%)	Relative standard deviation (%)
Radon				
Alpha track detectors	9	170 to 2000 Bq/m^3	18	24
Charcoal canisters	223	160 to 1900 Bq/m^3	12	18
Continuous monitors	88	220 to 1400 Bq/m^3	11	12
Grab sampling	60	420 to 3700 Bq/m^3	13	12
Electret-PERM	97	170 to 2000 Bq/m^3	18	20
Radon progeny				
Continuous working-level monitors	70	0.02 to 0.15 WL[d]	11	171[e]
Grab working-level sampling	48	0.01 to 0.21 WL	19	13
Radon progeny integrating sampling unit (RPISU)	4	0.04 to 0.07 WL	14	36

[a]Based on Gearo et al. (1988) and Research Triangle Institute (1988).
[b]Each set consists of 4 separate measurements.
[c]Median of the Absolute Relative Error
[d]1 Working level, WL, ($\approx 2.08 \times 10^{-5}$ J/M^3) = The potential alpha energy concentration of progeny in equilibrium with a radon concentration of 3700 Becquerels per cubic meter (Bq/m^3).
[e]Large standard deviation was primarily due to two sets of data with enormous absolute relative error.

Asbestos

OVERVIEW

Very few studies have been made of asbestos in homes. This section contains analytical methods used to measure airborne asbestos in the workplace and in post-abatement situations. The same methods can, in principle, be applied to asbestos in homes.

Because of their small size, airborne asbestos fibers are difficult to distinguish from other man-made and natural fibers. Fibers are usually collected on either a cellulose ester filter or a polycarbonate filter and measured visually under relatively high magnification, using phase contrast microscopy (PCM), scanning electron microscopy (SEM), or transmission electron microscopy (TEM). All three methods require the visual counting of fibers in randomly selected grid sections of slides prepared from the collected material. The key features of PCM, SEM, and TEM are shown in Table 4.

PHASE CONTRAST MICROSCOPY

The standard protocol for measuring exposure to airborne asbestos in the industrial workplace adopted by the Occupational Safety and Health Administration (OSHA) specifies PCM as the analytical method. The method (NIOSH 7400) specifies a 25-mm diameter, 0.8 to 1.2 μm pore size cellulose ester membrane filter for collection. Microscope slides are prepared from the collected material and the fibers in randomly chosen regions of each slide are counted using a positive phase contrast microscope with a 100 μm diameter graticule. One of two sets of counting rules are used to define which fibers are counted. The "A" Rules require that only fibers longer than 5 μm and with an aspect ratio (length to diameter) greater than or equal to 3:1 be counted. The "B" Rules require that only the ends of fibers longer than 5 μm and less than 3 μm in diameter and with an aspect ratio equal to or greater than 5:1 be counted. The final fiber count for the B Rules is determined by dividing the number of ends by 2. Fiber density on the filter is reported in fibers/mm^2; fiber concentration in the original air sample is reported in fibers/mL (Eller, 1984).

There are two serious limitations in the protocols described above. The first limitation is that PCM cannot distinguish between asbestos and non-asbestos fibers; all fibers or elongated

TABLE 4. COMPARISON OF PCM, SEM, AND TEM[a]

	PCM	SEM	TEM
Specificity for asbestos	Not specific (all fibers > 5 μm long are counted)	More specific but not definitive	Definitive (with options)
Magnification	400	1000-2000	20000
Sensitivity (thinnest fiber visible)	0.15 μm (best) 0.25 μm (typical)	0.05 μm (best) 0.20 μm (typical)	0.0002 μm (best) 0.0025 μm (typical)

[a]From U.S. Environmental Protection Agency (1985b). See text for definitions of PCM, SEM, TEM.

particles that meet the length, diameter, and aspect ratio criteria are counted. The second limitation is a result of the optical magnification used in PCM: Only fibers or particles 0.25 μm in diameter or larger can be seen (U.S. Environmental Protection Agency, 1985b).

SCANNING ELECTRON MICROSCOPY (SEM)

Scanning electron microscopy (SEM) is more sensitive to thin fibers and has a better specificity for asbestos than PCM. Fibers are typically collected on a 0.4 to 0.8 μm pore size polycarbonate (or cellulose ester) filter, carbon-coated directly on the filter, and transferred to an EM grid. The fiber substrate is relatively thick and the electrons bombarding the specimen during visual analysis are scattered and reflected rather than transmitted. These electrons are detected as noise by the microscope and restrict the visual range to fibers with a diameter of about 0.20 μm or larger (Environmental Protection Agency, 1985b).

At present, there is no standard method, no quality assurance laboratory testing, and no National Institute of Science and Technology (NIST, formerly NBS) reference materials for SEM. In spite of this, SEM is still more available and much less expensive than TEM.

TRANSMISSION ELECTRON MICROSCOPY (TEM)

Transmission electron microscopy (TEM) is considered the better of the two types of electron microscopy used for measuring airborne asbestos. There are two methods for collecting fibers and preparing slides for TEM analysis. One method requires the collection of fibers on a 0.4 μm pore size polycarbonate filter. A strip of the polycarbonate filter is carbon coated, placed on a TEM grid, and cleared by Jaffe washer or condensation washer. The second method requires the collection of fibers on a 0.45 μm pore size cellulose ester filter. A wedge of the cellulose ester filter is collapsed on a glass slide, etched in a low temperature plasma asher, carbon coated, and transferred to a TEM grid using a Jaffe washer. By both methods, the mounted fibers are identified as asbestos (from fiber morphology, selected area electron diffraction (SAED) patterns, and energy dispersive x-ray analysis (EDXA), measured, and counted at 20,000x magnification (U.S. Environmental Protection Agency, 1985b; Federal Register, 1987).

The major disadvantages of TEM are the cost and the time required for analysis. It is also less available than the other two methods. The analysis can be broken down into three levels to reduce cost and time for analysis: (1) identification of asbestos (for screening purposes), (2) elemental analysis of selected fibers (for regulatory action), and (3) quantitative analysis of a few representative fibers (for confirmatory analysis) (Yamate et al., 1984; Federal Register, 1987).

FIBROUS AEROSOL MONITOR (FAM-1)

The fibrous aerosol monitor (FAM-1) is a real-time direct reading instrument for measuring airborne fibers. The detectability of fibers depends on both length and diameter. For example, fibers 5 μm long can be detected down to a diameter of about 0.75 μm; fibers 10 μm long can be detected down to a diameter of about 0.6 μm (Lilienfeld, 1987). The FAM-1 induces fibers to oscillate by means of an electric field and detects the light scattering signature resulting from the oscillation of the fibers under illumination by a helium-neon laser beam (Monitoring Instruments for the Environment, no date). This approach allows the counting of fibers in the presence of nonfibrous particles but does not discriminate between asbestos and non-asbestos fibers.

Organic Compounds

OVERVIEW

A large number of organic compounds are typically present in indoor air. They range from compounds that are gases at ambient conditions to non-volatile compounds adsorbed onto suspended particulate material. Table 5 may be used to classify individual compounds on the basis of boiling point and provides information on typical sampling methods for each class. Most gas phase organic compounds may be measured by one or more of the general analytical methods described in the next section. More detailed information for specific compounds may be found in the literature (e.g., Riggin and Purdue, 1984; Riggin et al., 1986). Because of the emphasis placed upon them in the literature, formaldehyde, polycyclic aromatic hydrocarbons, and pesticides are covered in separate sections of this chapter.

GENERAL ANALYTICAL METHODS FOR GAS PHASE ORGANIC COMPOUNDS

Little is currently known about the concentrations and health risks of most organics in indoor air (Seifert and Ullrich, 1987). The Total Exposure Assessment Methodology (TEAM) Studies (Wallace, 1987; Pellizzari et al., 1987a,b; Handy et al., 1987), under the sponsorship of the U.S. EPA, represent the most in-depth studies of indoor pollutants to date. Personal air, fixed-site air, drinking water, and breath samples from individuals and homes in several states were analyzed for twenty selected organic chemicals, several of them carcinogenic. The methods used in the TEAM studies are rapidly becoming standard.

Emissions of specific organics from building materials and consumer products have been evaluated in the laboratory by various methods (Merrill et al., 1987; Wallace et al., 1987; Girman et al., 1987). Emissions from indoor combustion of fuels has also been studied (Traynor, 1987; Traynor et al., 1982; Spengler and Cohen, 1985).

As a rule, it is quite difficult to accurately measure a specific organic compound directly within the matrix of other components normally present in indoor air. The determination usually involves *collection* of a sample by some appropriate means, *separation* by gas chromatography, and *measurement* with an appropriate detector.

TABLE 5. CLASSIFICATION OF ORGANIC POLLUTANTS[a]

Description	Boiling Range[b] (°C)	Sampling Methods Used in Field Studies
Very Volatile Organic Compounds	<0 to 50-100	Batch sampling, adsorption on charcoal
Volatile Organ Compounds	50-100 to 240-260	Adsorption on tenax, carbon molecular black, or charcoal
Semi-Volatile Organic Compounds	240-260 to 380-400	Adsorption on polyurethane foam (PUF) or XAD-2
Particulate Organic Material[c]	>380	Collection on filters

[a]Adapted from World Health Organization (1987).
[b]Polar compounds tend to boil near the higher end of the range; less-polar compounds, near the lower end.
[c]Includes organic compounds associated with particulate matter.

Collection. Samples may be collected on a sorbent, in an impinger solution, or in an evacuated canister. Sorbent and impinger collection are more selective than canister collection although all methods have inherent disadvantages (Jayanty, 1989).

Ideally, a sorbent used for sample collection will have a strong affinity for the compound(s) of interest and little or no affinity for other species (H_2O, CO_2, etc.) in the matrix. Properties of several common sorbents are given in Table 6 (Sheldon et al., 1985; Raymer and Pellizzari, 1987; Levins, 1979; Krost et al., 1982; Piecewicz et al., 1979; Sanchez et al., 1987; Raymond and Guiochon, 1975; Riggin, 1984).

TABLE 6. PROPERTIES OF SOME COMMON SORBENTS

Sorbent	Properties
Tenax GC Resin	Porous polymer of 2,6-diphenyl-p-phenylene oxide; hydrophobic; not suitable for very light organics; lower capacity than XAD-2 but may be thermally desorbed.
XAD-2 Resin	A polystyrene-divinylbenzene porous polymer; hydrophobic; not suitable for very light organics; higher capacity than Tenax GC but requires solvent desorption.
Activated Charcoal	Typically a coconut charcoal; relatively high retention of water; organics are very strongly adsorbed; requires solvent desorption.
Graphitized Carbon Black	Obtained by heating thermal carbon blacks at 3000°C under an inert gas; nonselective; low retention of water and light gases; may be thermally desorbed.
Carbon Molecular Sieves	Pyrolyzed porous beads of polyvinylidene chloride; low capacity for water but organics are very strongly adsorbed; higher capacity for organics than graphitized carbon blacks; better than Tenax GC for highly volatile compounds; very light, volatile compounds may be thermally desorbed.

Sorbents offer the advantage of concentrating the sample as it is collected but require a desorption step prior to measurement. Desorption may be accomplished thermally or with a suitable solvent.

Impinger collection concentrates the analyte of interest through use of an absorbing solution. Impinger collection may also be used to stabilize very reactive substances, perhaps through derivatization.

Alternatively, samples may be collected in evacuated canisters. The inside walls of the canister must be chemically and physically inert to the species of interest. Stainless steel canisters with specially treated interior walls are available. Samples may also be collected in glass bulbs or bags made of an appropriate material (Jayanty, 1989).

Separation. Separation of organic gases and vapors is generally accomplished by gas chromatography (GC). Typically, an open-bore capillary column is used for the separation. The method is simple and relatively fast, and the species of interest can be measured as it exits the GC column by use of an appropriate detector. Cryogenic focusing or sorbent trapping may be used in the inlet of the GC to concentrate the sample and lower the detection limit for the species of interest.

Measurement. Gas chromatography detectors represent the most commonly used measurement method for airborne organics. Mass spectrometry, often associated with gas chromatography (GC-MS), is also used. A more recent approach involves the use of an array of sensors to selectively detect and measure a specific compound within the matrix (i.e., without separation).

Choice of an appropriate GC detector depends upon response to the compound(s) of interest, response to other species in the sample matrix, and desired sensitivity. Table 7 summarizes key features of some commonly used GC detectors.

Mass spectrometry (MS) provides one of the most powerful tools of chemical analysis. Every vaporizable compound gives a unique, often complex mass spectrum. Mass spectrometers used to measure components of complex mixtures must be used either (1) downstream from a gas chromatograph or (2) in conjunction with a sophisticated data system. Chemical ionization, rather than electron impact ionization, may be used to reduce fragmentation and simplify the mass spectrum, but the quantity of data acquired from a single analysis of a mixture is quite large.

An array of piezoelectric quartz crystals, each coated with a different partially selective material, may be used for direct analysis of multicomponent mixtures (Carey and Kowalski, 1986, 1988; Carey et al., 1987). The response of the array to a particular analyte resembles a typical absorption or emission spectrum with the resolution dependent on the number of sensors in the array. The pattern of the response is used to identify the analyte; analyte concentration is calculated from the magnitude of the response.

FORMALDEHYDE

Formaldehyde, because of its reactivity, is often collected in an impinger containing water. Formaldehyde reacts on contact with water to form methylene glycol. The water

TABLE 7. COMMON GAS CHROMATOGRAPHY DETECTORS

Detector	Detection Limit	Applications
Argon Ionization Detector	sub ppm to ppb	Responds to compounds with an ionization potential < 11.8 eV
Electron Capture Detector	pg	Responds to electron-capturing species, especially to halogenated compounds
Flame Ionization Detector	sub ppm sub ng	Responds to all combustible substances
Flame Photometric Detector	≥ sub ppb	Responds to compounds containing sulfur or phosphorus
Far Ultra-Violet Detector	low ng	Responds to all substances except the noble gases
Hall Electrolytic Conductivity Detector	50-100 pg	Responds to compounds containing a halogen, nitrogen or phosphorus
Helium Ionization Detector	sub ppm to ppb	Responds to compounds with an ionization potential < 19.8 eV
Mass Selective Detector	10 pg	Responds to all substances; may be used to identify as well as measure individual components of a mixture

TABLE 7 (cont'd). COMMON GAS CHROMATOGRAPHY DETECTORS

Detector	Detection Limit	Applications
Nitrogen Phosphorus Detector	10 pg	Responds to compounds containing nitrogen or phosphorus
Photo-Ionization Detector	10 pg $\geq$ ppb	Responds to compounds with an ionization potential < 11.7 eV
Redox Chemiluminescence Detector	sub ng	Responds to compounds containing oxygen, sulfur, nitrogen or phosphorus and to unsaturated hydrocarbons
Thermal Conductivity Detector	< 100 ppm	Responds to all compounds

solution is then treated with either chromotropic acid or pararosaniline and the resulting solution analyzed spectrophotometrically (National Research Council, 1981; Hawthorne and Matthews, 1987; Georghiou et al., 1987).

Diffusion badges containing a water solution may also be used to sample formaldehyde in air. After exposure, the badge solution is analyzed colorimetrically. A comparison of impinger and diffusion badge collection (Stock, 1987) indicates close agreement of the two sampling methods.

Silica gel coated with acidified 2,4-dinitrophenylhydrazine may be used to sample aldehydes and ketones in air (Tejada, 1986). Analysis of the sample, which contains the hydrazone derivatives of any aldehydes and ketones collected, is typically accomplished by high performance liquid chromatography (HPLC).

POLYCYCLIC AROMATIC HYDROCARBONS

Polycyclic aromatic hydrocarbons (PAHs) in air may (1) exist as a vapor, (2) be adsorbed onto atmospheric particulate material, or (3) be distributed between the vapor and adsorbed states. Sampling usually involves the collection of particulate-bound material on a filter and the collection of vapors on an adsorbent or in an impinger (Davis et al., 1987, and Otson et al., 1987).

The particulate-bound PAHs are typically extracted with a liquid solvent or a supercritical fluid (Hawthorne and Miller, 1987). Recovery is complicated by the vaporization of loosely bound PAHs from the particulate material after collection and by the difficulty of desorbing very tightly bound PAHs (Engelbach et al., 1987). Spitzer (1982) has shown that XAD-2 resin will separate PAHs from polar and non-polar contaminants in air particulate material. Extracts are usually analyzed by ultraviolet absorptiometry, fluorescence spectrometry, or gas chromatography/mass spectrometry, but a number of other methods have also been reported (Davis et al., 1987).

PESTICIDES

Carcinogenic species used in pesticides and frequently found in indoor air include chlordane, heptachlor/heptachlor epoxide, aldrin, and dieldrin (Cavender et al., 1986 and 1987). These and similar species are typically collected on polyurethane foam (PUF) or PUFin combination with some granular sorbent (e.g., XAD-2, Tenax GC, or Florisil PR). Following Soxhlet extraction and concentration, the samples are analyzed by GC-ECD, GC-MS, or GC-MS-MID (multiple ion detection) (Hsu et al., 1988; Lewis and Jackson, 1982; Lewis and MacLeod, 1982; Lewis et al., 1986).

Inorganic Species

OVERVIEW

In addition to asbestos, a number of inorganic pollutants, especially some of the heavy metals, have been classified as carcinogens. The inorganic species of most concern are inorganic arsenic (salts, arsenates, and arsenites); beryllium; cadmium (oxide, bromide, and chloride); chromium (hexavalent); nickel (carbonyl and subsulfide); and selenium (sulfide).

Table 8 summarizes sampling and analysis approaches that are currently used to measure concentrations of individual metals in workplace atmospheres. They should also be appropriate for indoor air. In most instances, more accurate methods of collection and analysis are available, but they have not yet been accepted as standard methods.

INORGANIC ARSENIC (SALTS, ARSENATES, AND ARSENITES)

National Institute for Occupational Safety and Health Method 7300 (Eller, 1984) is appropriate for 29 elements including arsenic. By this method, the sample is collected on a cellulose ester membrane filter and, after preparation, analyzed by inductively coupled argon plasma, atomic emission spectroscopy.

NIOSH Method 7900 is designed to measure arsenic and its compounds as arsenic. By this method, the sample is again collected on a cellulose ester membrane but, after preparation, is analyzed by atomic absorption, flame arsine generation. The estimated limit of detection for this method is 0.02 μg per sample (Eller, 1984).

BERYLLIUM

Beryllium occurs in air as a component of suspended particulate matter and may be collected on a membrane filter. Analysis by gas chromatography (Ross and Sievers, 1972) or atomic absorption spectroscopy (AAS) (Owens and Gladney, 1975) appears to offer adequate sensitivity, but both require pretreatment to remove interfering species (U.S. Environmental Protection Agency, 1987). Beryllium may also be measured by inductively coupled plasma emission.

310

**TABLE 8. SAMPLING AND ANALYSIS APPROACHES
FOR INORGANIC SPECIES**

Species	Sampling	Analysis	LOD	Ref.
Arsenic (salts, arsenites, arsenates)	Cellulose ester membrane	ICAP-AES		a
		AA-Flame Arsine Generation	0.02 μg	a
	Membrane filter	GC		d,e
		AAS		d,f
Beryllium	Cellulose ester membrane	ICAP-AES		a
		AA-Graphite Furnace	0.05 μg	a
Cadmium (oxide, bromide, chloride)	Cellulose ester membrane	ICAP-AES		a
		AAF		a
		IC		b
	Filter	Effect on Oxid. of I^- by H_2O_2	0.001 μg	g
Chromium(VI)	Cellulose ester membrane	ICAP-AES		a
	PVC filter	VAS	0.05 μg	a
Nickel carbonyl	Impinger	Chemiluminescence	ppb	h
	Filter	AAF	0.005 μg/mL	i,j
Nickel subsulfide		AAS, XRF, ICAP, colorimetry, SSMS, NAA, FES		c
	Cellulose ester membrane	ICAP-AES	1 μg	a
Selenium sulfide	Cellulose ester membrane	ICAP-AES	1 μg	a

[a]Eller (1984)

[b]U.S. Environmental Protection Agency (1984)

[c]U.S. Environmental Protection Agency (1985a)

[d]U.S. Environmental Protection Agency (1987b)

[e]Ross and Sievers (1972)

[f]Owens and Gladney (1975)

[g]Kneebone and Freiser (1975)

[h]Stedman et al. (1979)

[i]Pickett and Koirtyohann (1969)

[j]Sachdev and West (1970)

Beryllium concentration may also be determined by NIOSH Method 7300, described above for arsenic, or by NIOSH Method 7102. Method 7102 requires collection on a cellulose ester membrane filter and, after sample preparation, analysis by atomic absorption, graphite furnace. The estimated limit of detection is 0.005 μg per sample (Eller, 1984).

CADMIUM (OXIDE, BROMIDE, AND CHLORIDE)

Cadmium may be analyzed by NIOSH Method 7300, described above for arsenic, or by NIOSH Method 7048. By Method 7048, the sample is collected on a cellulose ester membrane filter and, after preparation, analyzed by atomic absorption with flame. The estimated limit of detection by this method is 0.05 μg per sample (Eller, 1984).

CHROMIUM (HEXAVALENT)

Chromium occurs in air as a component of suspended particulate matter and may be collected on a filter. Filters are typically composed of cellulose, polyethylene, polystyrene, PVC, or glass (U.S. Environmental Protection Agency, 1984). Chromium(VI) may be separated from matrix materials by ion chromatography.

Chromium(VI), collected on filters, may be measured by a catalytic method developed by Kneebone and Freiser (1975). The measurement is based on the catalytic effect of chromium(VI) on the oxidation of iodide by hydrogen peroxide. The sensitivity of the method is 0.001 μg Cr(VI).

Chromium may also be measured by NIOSH Method 7300 as described previously for arsenic. Hexavalent chromium may be measured using NIOSH Method 7600 which requires sample collection on a PVC membrane filter and, after preparation, analysis by visible absorption spectrophotometry. The estimated limit of detection for this method is 0.05 μg per sample (Eller, 1984).

NICKEL (CARBONYL AND SUBSULFIDE)

Trace amounts of nickel in ambient air are almost always present as a component of suspended particulate matter. As a result, nickel is usually collected, along with other particulate material, on a filter. Unfortunately, nickel carbonyl, because of its volatility, cannot be collected in this fashion. In sampling nickel from flue gases, an impinger is typically used to trap gaseous species (U.S. Environmental Protection Agency, 1985a).

The most commonly used analytical method for nickel in air is atomic absorption spectrophotometry with flame (AAF). The detection limit for nickel by this method is 0.005 μg/mL (Pickett and Koirtyohann, 1969; Sachdev and West, 1970). Other methods include atomic absorption spectrophotometry without flame, x-ray fluorescence spectrometry (XRF), inductively coupled argon plasma (ICAP) spectroscopy, colorimetry, spark source mass spectrometry (SSMS), neutron activation analysis (NAA), and flame emission spectrophotometry (FES) (U.S. Environmental Protection Agency, 1985a).

Analysis for specific nickel compounds in ambient air is quite difficult due to induced chemical changes inherent in the analytical techniques and to the very low concentrations of nickel compounds in ambient air. A variety of techniques for identifying compound form have been attempted on flyash samples (U.S. Environmental Protection Agency, 1985a).

Vapor-phase inorganic nickel compounds (e.g., nickel carbonyl) are easily separated from other nickel compounds on the basis of their volatility. The chemiluminescence analytical method is quite specific for nickel carbonyl and has a detection limit in the parts-per-billion range (Stedman et al., 1979).

Nickel may also be measured by NIOSH Method 7300 as described previously for arsenic. The estimated instrumental limit of detection for nickel is 1 μg per sample (Eller, 1984).

SELENIUM (SULFIDE)

Selenium may be measured by NIOSH Method 7300 as described previously for arsenic. The estimated instrumental limit of detection for selenium is 1 μg per sample (Eller, 1984).

Particles

OVERVIEW

Particulate matter is made up of solid aerosols (e.g., dusts and smokes) and liquid aerosols (e.g., mists and fogs). Table 9 gives brief descriptions of devices used to collect and measure particles. Environmental tobacco smoke is perhaps the most notorious of the pollutants containing particulate material that may be present in indoor air.

ENVIRONMENTAL TOBACCO SMOKE

Environmental tobacco smoke (ETS) is a complex mixture of gaseous substances and suspended particulates and is quite difficult to measure in indoor air. More than 4,500 compounds have been identified in tobacco smoke: Some exist completely in the vapor phase, others are adsorbed onto particulate material, and still others are distributed between the vapor and adsorbed phases. As a result of adsorption onto and desorption from suspended particles, and even some chemical changes in the more reactive compounds, the components of ETS show a pronounced spatial and temporal distribution in indoor environments. In addition, background indoor air is a complex mixture containing a number of the chemical species also found in ETS. Even though body burden of specific components of ETS can be determined, there are no direct measures of total dose. However, there are several methods for assessing exposure.

Respirable suspended particle concentrations have been determined using personal monitors in order to estimate exposure to ETS (Spengler et al., 1985). In addition, personal and indoor space monitoring have been used to measure concentrations of specific compounds or classes of compounds and these (e.g., nicotine) may be used to indicate exposure to ETS (Hammond et al., 1987; Muramatsu et al., 1984). Two recent studies on the chemical composition of ETS suggests potential gas-phase (Eatough et al., 1989) and particulate phase (Benner et al., 1989) tracers based on environmental chamber experiments.

Sampling of ETS components may be active or passive. Active samplers utilize filters and/or vapor traps to collect material; passive samplers utilize diffusion and permeation to concentrate collected gases and vapors. In both cases, the samples are subsequently analyzed in the laboratory. Particles are typically measured by light-scattering principles or frequency

314

changes induced in piezoelectric quartz crystals. Gases are typically measured using infrared absorption, electrochemical reactions, or gas chromatography with appropriate detection (National Research Council, 1986; Hammond et al., 1987). Studies found in the literature (described in Guerin et al., 1987) suggest that determination of a specific compound (nicotine, for example) in ETS should include measurements of the quantity/concentration of the compound in both particle-bound and vapor- phase material (Hammond et al., 1987) since, as cigarette smoke ages some compounds change from particulate to vapor phase and vice versa.

Other approaches used to assess exposure to ETS include measuring air and body nicotine and cotinine, and nitrosamines: questionnaires; and exposure modeling. These topics are described in a later section of this work.

TABLE 9. SAMPLING DEVICES AND ANALYSIS METHODS FOR PARTICLES[a]

Device or Method	Description
Sampling Devices:	
settling chamber	Particles from a trapped sample of air settle by gravity onto a microscope slide.
centrifugal device	Centrifugal force, inside a small cyclone or curved surface trap, is used to separate particles on the basis of size.
impinger	A glass nozzle submerged in a liquid is used to collect particles.
impactor	Particles are collected from an aerosol stream by impaction onto a surface.
filter	Particles are collected by passing an air sample through a filter.
electrostatic precipitator	An ionizing electrode and a collection (or grounded) electrode are used to collect particles.
thermal precipitator	Particles are removed from an aerosol on the basis of response to a temperature gradient.
Analysis Methods:	
microscopy	Allows previously collected dust particles to be counted, sized, and/or identified.
piezoelectric	Allows real-time monitoring of particulate mass concentration by deposition of particles onto a quartz crystal.
optical	Allows a direct measurement of particles in air on the basis of particle interactions with light.
electrical	Allows direct measurement of particles in air on the basis of the tendency of airborne particles to acquire electrical charge.
beta attenuation	The collected mass of airborne particles is measured by its attenuation of beta radiation passing through it.

[a]Based on Lioy and Lioy (1983).

Nonionizing Radiation

OVERVIEW

Non-ionizing radiation (NIR) refers to electromagnetic radiation with wavelengths longer than 100 nm. It includes ultra-violet (UV); visible; infrared (IR); radiofrequency (RF), including microwaves (MW); and extremely low frequency (ELF) fields, including power frequencies of 50-60 Hz. Pressure waves (e.g., infrasound and ultrasound) are also included (International Radiation Protection Association, 1985).

All electromagnetic radiation consists of an electric field and a magnetic field. Either or both, or their effects on living tissues, may be measured. The analytical approach used to measure NIR depends upon the purpose of the measurement. Radiometry deals with quantities (e.g., field intensity) associated with radiation fields; interaction coefficients deal with the interaction of radiation and matter; and dosimetry deals with quantities used to determine the specific absorption rate (SAR) of biological tissue (International Radiation Protection Association, 1985).

RADIOMETRIC MEASUREMENTS

Electric and magnetic fields associated with radiation of a particular wavelength oscillate at the same frequency, typically 50 or 60 cycles per second from alternating-current (AC) power lines. The fields have both a magnitude and a direction. The field intensity, or field strength, of an electric field is typically measured in volts per meter; the field intensity of a magnetic field, in teslas or gauss (Shepard et al., 1987).

The Electric Power Research Institute (EPRI) has funded the development of EMDEX (Electric and Magnetic Field Digital Exposure), a compact, lightweight instrument used to monitor personal exposure to electromagnetic radiation. Exposure assessments are obtained by combining the field intensity measurements and computer models which can extrapolate measured values to a wide variety of environments (Sussman, 1988; Shepard et al., 1987).

Cahill and Elder (1984) have described several other instruments, both developmental and commercial, for measuring the intensity of electric and magnetic fields.

INTERACTION COEFFICIENTS

Interactions of NIR with matter include attenuation, absorption, scattering, and reflection and related phenomena. An interaction coefficient may be calculated for each type of interaction (e.g., a scattering coefficient). The term "coefficient" generally refers to the relative decrease of a radiometric quantity due to an interaction phenomenon during passage through a thin layer of medium divided by the thickness of the layer. Interaction coefficients are expressed in reciprocal meters (m^{-1}) (International Radiation Protection Association, 1985).

DOSIMETRY

Dosimetry is the quantification of the specific absorption rate (SAR) of a biological entity. The SAR is dependent on the configuration of the source, the physical characteristics of the exposed subject, the orientation of the exposed subject with respect to the source, and the frequency of the electromagnetic radiation. Determination of the SAR of tissues requires measurements with electric field probes, thermocouples, thermistors, fiber optic probes, thermography, and calorimetry. The appropriate method for a particular application depends upon the frequency range of the radiation, the type of subject or biological preparation, and whether distributive or whole-body average SAR is desired (Guy, 1987).

Other Approaches to Assessing Exposure

BIOLOGICAL MARKERS

The assessment of exposure to some pollutants may be made by analysis of physiological fluids. Biochemical methods may be used to obtain estimates of exposure based on the uptake of specific agents in body fluids.

Biological markers have been used to estimate exposure to environmental tobacco smoke (ETS). For example, The presence of nicotine and its major metabolite, cotinine, in biological fluids is entirely due to exposure to tobacco, tobacco smoke, or environmental tobacco smoke. The determination of nicotine and cotinine in the saliva, blood, or urine of active or passive smokers is done primarily by gas chromatography (GC) with a nitrogen-sensitive detector, or by radioimmunoassay (RIA). The GC method can be used to measure concentrations of nicotine as low as 1 ng/mL and concentrations of cotinine as low as 6 ng/mL in biological fluids. The radioimmunoassays for nicotine and cotinine represent a newer analytical method. The sensitivity of these assays is about 0.5 ng/mL for both nicotine and cotinine and has inter- and intra-assay variation of $\pm 5\%$ (Langone et al., 1973; Hill et al., 1983). The RIA method has been used by only a limited number of laboratories because it requires the synthesis of specific nicotine and cotinine derivatives for the generation of serum albumin conjugates and the raising of antibodies to these conjugates (Langone et al., 1973). The RIA method also requires careful drawing and handling of samples to avoid contamination.

Cotinine, the major metabolite of nicotine, offers several advantages as a biological marker for ETS exposure. Cotinine is specific for tobacco and its quantitation can be useful in gathering information from large populations of both smokers and non-smokers.

The use of cotinine as an indicator of side-stream smoke exposure in children has been studied by Greenberg et al. (1984). The study revealed a high correlation between the exposure of children at home to side-stream smoke and the levels of cotinine in their urine.

QUESTIONNAIRES

Questionnaires can be useful for the assessment of exposure to some pollutants (e.g., ETS). A questionnaire may be used, for example, to determine the physical characteristics of microenvironments within a home and the activity patterns of those living there. From this information, individuals may be classified with regard to broad categories of exposure (National Research Council, 1986). It is not a simple matter, however, to design questions which will elicit unambiguous replies and permit quantitative estimates of individual exposures.

EXPOSURE MODELING

Exposure modeling requires knowledge of the concentrations of air contaminants in all the microenvironments within a residence or work area and the time individuals spend in each of those microenvironments. Typically, a model must take into account generation rate, ventilation, infiltration, mixing, removal by adsorption onto surfaces, and volume of space in which exposure occurs. Models are particularly useful in making exposure estimates in situations where measured concentrations are not available.

References

Benner, C. L., J. M. Bayona, F. M. Caka, H. Tang, L. Lewis, J. Crawford, J. D. Lamb, M. L. Lee, E. A. Lewis, L. D. Hansen, and D. J. Eatough, (1989), "Chemical Composition of Environmental Tobacco Smoke. 2. Particulate-Phase Compounds," Environ. Sci. Technol., Vol. 23, No. 6:688-699.

Cahill, D. F., and J. A. Elder, eds., (1984), Biological Effects of Radiofrequency Radiation, Report No. EPA-600/8-83-026F, Health Effects Research Laboratory, Office of Research and Development, U.S. Environmental Protection Agency, Research Triangle Park, NC 27711.

Carey, W. P., K. R. Beebe, and B. R. Kowalski, (1987), "Multicomponent Analysis Using an Array of Piezoelectric Crystal Sensors," Anal. Chem., Vol. 59, No. 11:1529-1534.

Carey, W. P., and B. R. Kowalski, (1986), "Chemical Piezoelectric Sensor and Sensor Array Characterization," Anal. Chem., Vol. 58, No. 14:3077-3084.

Carey, W. P., and B. R. Kowalski, (1988), "Monitoring a Dryer Operation Using an Array of Piezoelectric Crystals," Anal. Chem., Vol. 60, No. 6:541- 544.

Cothern, C. R., and J. E. Smith, Jr., eds., (1987), Environmental Radon, Plenum Press, New York.

Davis, C. S., P. Fellin, and R. Otson, (1987), "A Review of Sampling Methods for Polyaromatic Hydrocarbons in Air," J. Air Pollut. Control Assoc., Vol. 37, No. 12:1397-1408.

Eatough, D. J., C. L. Benner, J. M. Bayona, G. Richards, J. D. Lamb, M. L. Lee, E. A. Lewis, and L. D. Hansen, (1989), "Chemical Composition of Environmental Tobacco Smoke. 1. Gas-Phase Acids and Bases," Environ. Sci. Technol., Vol. 23, No. 6:679-687.

Eller, P. M., ed., (1984), NIOSH Manual of Analytical Methods, 3rd Ed., Cincinnati, OH., U.S. Department of Health and Human Services, National Institute for Occupational Safety and Health.

Engelbach, R. J., A. A. Garrison, E. L. Wehry, and G. Mamantov, (1987), "Measurement of Vapor Deposition and Extraction Recovery of Polycyclic Aromatic Hydrocarbons Adsorbed on Particulate Solids," Anal. Chem., Vol. 59, No. 20:2541-2543.

Gearo, J. R., Jr, H. M. Singletary, and D. F. Naugle, (1988), "The Growth of the National Radon Measurement Proficiency (RMP) Program," The 1988 Symp. on Radon and Radon Reduction Technol., Vol. II. EPA/600/9-89/006b. Research Triangle Park, NC, Environmental Protection Agency, Air & Energy Engineering Research Laboratory, 1-1 to 1-14.

321

Georghiou, P. E., L. Winsor, C. J. Shirtliffe, and J. Svec, (1987), "Storage Stability of Formaldehyde Solutions Containing Pararosaniline Reagent," Anal. Chem., Vol. 59, No. 19:2432-2435.

Girman, J. R., and A. T. Hodgson, (1987), "Considerations in Evaluating Emissions from Consumer Products," Atmospheric Environment, Vol. 21, No. 2:315-320.

Greenberg, R. A., N. J. Haley, R. A. Etzel, and F. A. Loda, (1984), "Measuring the Exposure of Infants to Tobacco Smoke: Nicotine and Cotinine in Urine and Saliva," N. Engl. J. Med., Vol. 310:1075-1078.

Guerin, M. R., C. E. Higgins, and R. A. Jenkins, (1987), "Measuring Environmental Emissions from Tobacco Combustion: Sidestream Cigarette Smoke Literature Review," Atmospheric Environment, Vol. 21, No. 2:291-297.

Guy, A. W., (1987), "Dosimetry Associated with Exposure to Non-Ionizing Radiation: Very Low Frequency to Microwaves," Health Physics, Vol. 53, No. 6:569-584.

Hammond, S. K., B. P. Leaderer, A. C. Roche, and M. Schenker, (1987), "Collection and Analysis of Nicotine as a Marker for Environmental Tobacco Smoke," Atmospheric Environment, Vol. 21, No. 2:457-462.

Handy, R. W., D. J. Smith, N. P. Castillo, C. M. Sparacino, K. Thomas, D. Whitaker, J. Keever, P. A. Blau, L. S. Sheldon, K. A. Brady, R. L. Porch, J. T. Bursey, E. D. Pellizzari, and L. Wallace, (1986), Standard Operating Procedures Employed in Support of an Exposure Assessment Study, Volume IV; Air, Toxics, and Radiation Monitoring Research Division; Office of Monitoring, System and Quality Assurance; U.S. Environmental Protection Agency, Office of Research and Development, Washington, DC 20460.

Hawthorne, A. R., and T. G. Matthews, (1987), "Models for Estimating Organic Emissions from Building Materials: Formaldehyde Example," Atmospheric Environment, Vol. 21, No. 2:419-424.

Hawthorne, S. B., and D. J. Miller, (1987), "Extraction and Recovery of Polycyclic Aromatic Hydrocarbons from Environmental Solids Using Supercritical Fluids," Anal. Chem., Vol. 59, No. 13:1705-1708.

Henschel, D. B., (1988), Radon Reduction Techniques for Detached Houses: Technical Guidance (Second Edition), EPA/625/5-87/019, Air and Energy Engineering Research Laboratory, Office of Environmental Engineering and Technology Demonstration, Office of Research and Development, U.S. Environmental Protection Agency, Research Triangle Park, North Carolina 27711; Revised.

Hill, P., N. J. Haley, and E. L. Wynder, (1983), "Cigarette Smoking: Carboxyhemoglobin, Plasma Nicotine, Cotinine and Thiocyanate Levels vs. Self-reported Data and Cardiovascular Disease," J. Chron. Dis., Vol. 36:439-449.

Hsu, J. P., Wheeler, H.G., Jr., D. E. Camann, H. J. Schattenberg, III, R. G. Lewis, and A. E. Bond, (1988), "Analytical Methods for Detection of Nonoccupational Exposure to Pesticides," J. Chromatogr. Sci., Vol. 26:181- 189.

International Radiation Protection Association (1985), "Review of Concepts, Quantities, Units and Terminology for Non-Ionizing Radiation Protection: A report of the International Non-Ionizing Radiation Committee of the International Radiation Protection Association," Health Physics, Vol. 49, No. 6.

Jayanty, R. K. M., (1989), "Evaluation of Sampling and Analytical Methods for Monitoring Toxic Organics in Air," Atmospheric Environment, Vol. 23, No. 4:777-782.

Kneebone, B. M., and H. Freiser, (1975), "Determination of Chromium(VI) in Industrial Atmospheres by a Catalytic Method," Anal. Chem., Vol. 47, No. 3:595-598.

Krost, K. J., E. D. Pellizzari, S. G. Walburn, and S. A. Hubbard, (1982), "Collection and Analysis of Hazardous Organic Emissions," Anal. Chem., Vol. 54, No. 4:810-817.

Kusnetz, H. L., (1956), "Radon Daughters in Mine Atmospheres--A Field Method for Determining Concentrations," Am. Ind. Hyg. Assoc. J., Vol. 17:85-88.

Langone, J. J., H. Gjika, and H. Van Vunakis, (1973), "Nicotine and Its Metabolites. Radioimmunoassays for Nicotine and Cotinine," Biochemistry, Vol. 12:5025-5030.

Levins, P. L., (1979), "The Use of Sorbent Resins in Environmental Sampling," Process Measurements Review, Vol. 1, No. 4:7-8, Spring Edition.

Lewis, R. G., and K. E. MacLeod, (1982), "Portable Sampler for Pesticides and Semivolatile Industrial Organic Chemicals in Air," Anal. Chem., Vol. 54, No. 2:310-315.

Lewis, R. G., and M. D. Jackson, (1982), "Modification and Evaluation of a High-Volume Air Sampler for Pesticides and Semivolatile Industrial Organic Chemicals," Anal. Chem., Vol. 54, No. 3:592-594.

Lewis, R. G., A. E. Bond, T. R. Fitz-Simons, D. E. Johnson, and J. P. Hsu, (1986), "Monitoring for Non-Occupational Exposure to Pesticides in Indoor and Personal Respiratory Air," Presented at the 79th Annual Meeting of the Air Pollution Control Association, Minneapolis, Minnesota, June 22-27, 1986, paper no. 86-37.4..

Lilienfeld, P., (1987), Asbestos Monitoring with the MIE Fibrous Aerosol Monitor, Monitoring Instruments for the Environment, Inc., 213 Burlington Road, Bedford, Massachusetts 01730.

Lioy, P. J., and M. J. Y. Lioy, eds., (1983), Air Sampling Instruments for Evaluation of Atmospheric Contaminants, 6th ed., American Conference of Governmental Industrial Hygienists, Cincinnati, Ohio.

Merrill, R. G., R. S. Steiber, R. F. Martz, and L. H. Nelms, (1987), "Screening Methods for the Identification of Organic Emissions from Indoor Air Pollution Sources," Atmospheric Environment, Vol. 21, No. 2:331-336.

Monitoring Instruments for the Environment, Inc. (MIE) (no date), Fibrous Aerosol Monitor, MIE, Inc., 213 Burlington Road, Bedford, Massachusetts 01730.

Muramatsu, M., S. Umemura, T. Okada, and H. Tomita, (1984), "Estimation of Personal Exposure to Tobacco Smoke with a Newly Developed Nicotine Personal Monitor," Environ. Res., Vol. 35:218-227.

National Research Council, (1986), Environmental Tobacco Smoke: Measuring Exposures and Assessing Health Effects, prepared by the Committee on Passive Smoking, Board on Environmental Studies and Toxicology, National Research Council; National Academy Press, Washington, D.C.

National Research Council, (1981), Formaldehyde and Other Aldehydes, Committee on Aldehydes, Board on Toxicology and Environmental Health Hazards, Assembly of Life Sciences, National Research Council; National Academy Press, Washington, D.C.

Otson, R., J. M. Leach, and L. T. K. Chung, (1987), "Sampling of Airborne Polycyclic Aromatic Hydrocarbons," Anal. Chem., Vol. 59, No. 13:1701-1705.

Owens, J. W., and E. S. Gladney, (1975), "Determination of Beryllium in Environmental Materials by Flameless Atomic Absorption Spectroscopy," At. Absorpt. Newsl., Vol. 14:76-77.

Pellizzari, E. D., K. Perritt, T. D. Hartwell, L. C. Michael, C. M. Sparacino, L. S. Sheldon, R. Whitmore, C. Leninger, H. Zelon, R. W. Handy, D. Smith, and L. Wallace, (1987a), Total Exposure Assessment Methodology (TEAM) Study: Elizabeth and Bayonne, New Jersey, Devils Lake, North Dakota, and Greensboro, North Carolina, Volume II, Final Report; Air, Toxics, and Radiation Monitoring Research Division; Office of Monitoring, System and Quality Assurance; U.S. Environmental Protection Agency, Office of Research and Development, Washington, DC 20460.

Pellizzari, E. D., K. Perritt, T. D. Hartwell, L. C. Michael, R. Whitmore, R. W. Handy, D. Smith, H. Zelon, and L. Wallace, (1987b), Total Exposure Assessment Methodology (TEAM Study): Selected Communities in Northern and Southern California, Volume III, Final Report; Air, Toxics, and Radiation Monitoring Research Division; Office of Monitoring, System and Quality Assurance; U.S. Environmental Protection Agency, Office of Research and Development, Washington, DC 20460.

Pickett, E. E., and S. R. Koirtyohann, (1969), "Emission Flame Photometry-- A New Look at an Old Method," Anal. Chem., Vol. 41:28A-42A.

Piecewicz, J. F., J. C. Harris, and P. L. Levins, Further Characterization of Sorbents for Environmental Sampling, Report No. EPA-600/7-79-216, sponsored by EPA, Office of Research and Development, Industrial Environmental Research Laboratory, Research Triangle Park, NC 27711.

Raymer, J. H., and E. D. Pellizzari, (1987), "Toxic Organic Compound Recoveries from 2,6-Diphenyl-p-phenylene Oxide Porous Polymer Using Supercritical Carbon Dioxide and Thermal Desorption Methods," Anal. Chem., Vol. 59, No. 7:1043-1048.

Raymond, A., and G. Guiochon, (1975), "The Use of Graphitized Carbon Black as a Trapping Material for Organic Compounds in Light Gases Before a Gas Chromatographic Analysis," J. of Chrom. Science, Vol. 13:173-177.

Research Triangle Institute (1988) National Radon Measurement Proficiency (RMP) Program, Analytical Proficiency Report, Round 5 Performance Test. Contract number 68-01-7350, U.S. Environmental Protection Agency

Riggin, R. M., (1984), "Evaluation of Carbon Molecular Sieves as Adsorbents for the Determination of Volatile Organic Compounds," presented at the APCA/ASQC Specialty Conference On: Quality Assurance in Air Pollution Measurements, Boulder, Colorado, October 14-18, 1984. Co-Sponsored by The Air Pollution Control Association and The American Society for Quality Control.

Riggin, R. M., and L. J. Purdue, (1984), Compendium of Methods for the Determination of Toxic Organic Compounds in Ambient Air, EPA-600/4-84-041; Quality Assurance Division, Environmental Monitoring Systems Laboratory, U.S. Environmental Protection Agency, Research Triangle Park, North Carolina 27711.

Riggin, R. M., W. T. Winberry, Jr., N. V. Tilley, and L. J. Purdue, (1986), Supplement to EPA/600/4-84/041: Compendium of Methods for the Determination of Toxic Organic Compounds in Ambient Air, EPA/600/4-87-006; Quality Assurance Division, Environmental Monitoring Systems Laboratory, U.S. Environmental Protection Agency, Research Triangle Park, North Carolina 27711.

Ross, W. D., and R. E. Sievers, (1972), "Environmental Air Analysis for Ultratrace Concentrations of Beryllium by Gas Chromatography," Environ. Sci. Technol., Vol. 6(2):155-158.
Sachdev, S. I., and P. W. West, (1970), "Concentration of Trace Metals by Solvent Extraction and Their Determination by Atomic Absorption Spectrophotometry," Environ. Sci. Technol., Vol. 4(9):749-751.

Samuelsson, C. (1988) "Retrospective Determination of Radon in Houses," Nature, Vol. 334:338-340.

Sanchez, D. C., M. Mason, and C. Norris, (1987), "Methods and Results of Characterization of Organic Emissions from an Indoor Material," Atmospheric Environment, Vol. 21, No. 2:337-345.

Seifert, B., and D. Ullrich, (1987), "Methodologies for Evaluating Sources of Volatile Organic Chemicals (VOC) in Homes," Atmospheric Environment, Vol. 21, No. 2:395-404.

Sheldon, L. S., C. M. Sparacino, and E. D. Pellizzari, (1984), "Review of Analytical Methods for Volatile Organic Compounds in the Indoor Environment," Indoor Air and Human Health, Proceedings of the Seventh Life Sciences Symposium, Knoxville, Tennessee, October 29-31, 1984, Chapter 26, pp. 335-349.

Shepard, M., L. Sagan, R. Black, S. Sussman, and C. Rafferty, (1987), "EMF: The Debate on Health Effects," EPRI Journal, pp. 4-15, October/November 1987.

Singletary, H. M., (1988), Coordinator, Natural Radionuclide Programs, Research Triangle Institute, Research Triangle Park, NC 27709; unpublished summary.

Spengler, J. D., R. D. Treitman, T. D. Tosteson, D. T. Mage, and M. L. Soczek, (1985), "Personal Exposures to Respirable Particulates and Implications for Air Pollution Epidemiology," Environ. Sci. Technol., Vol. 19:700-707.

Spengler, J. D., and M. A. Cohen, (1985), "Emissions from Indoor Combustion Sources," Indoor Air and Human Health, Proceedings of the Seventh Life Sciences Symposium, Knoxville, Tennessee, October 29-31, 1984, Chapter 20, pp. 261-278.

Spitzer, T., (1982), "Clean-Up of Polynuclear Aromatic Hydrocarbons from Air Particulate Matter on XAD-2," J. of Chrom., No. 237:273-278.

Stedman, D. H., D. A. Tammaro, D. K. Branch, and R. Pearson, Jr. (1979), "Chemiluminescence Detector for the Measurement of Nickel Carbonyl in Air," Anal. Chem., Vol. 51:2340-2342.

Stock, T. H., (1987), "Formaldehyde Concentrations Inside Conventional Housing," J. Air Pollut. Control Assoc., Vol. 37, No. 8:913-918.

Sussman, S. S., (1988), "Electric and Magnetic Field Exposure Assessment," EPRI Journal, pp. 50-52.

Tejada, S. B., (1986), "Evaluation of Silica Gel Cartridges Coated In Situ with Acidified 2,4-Dinitrophenylhydrazine for Sampling Aldehydes and Ketones in Air," Intern. J. Environ. Anal. Chem., Vol. 26, pp.167-185.

Traynor, G. W., M. G. Apte, J. F. Dillworth, C. D. Hollowell, and E. M. Sterling, (1982), "The Effects of Ventilation on Residential Air Pollution Due to Emissions from a Gas-Fired Range," Environment International, Vol. 8:447-452.

Traynor, G. W., (1987), "Field Monitoring Design Considerations for Assessing Indoor Exposures to Combustion Pollutants," Atmospheric Environment, Vol. 21, No. 2:377-383.

Tsviglou, E. G., H. E. Ayers, and D. A. Holaday, (1953), "Occurrence of Nonequilibrium Atmospheric Mixtures of Radon and Daughters," Nucleonics, Vol. 11:40.

U.S. Environmental Protection Agency, (1984), Health Assessment Document for Chromium, EPA/600/8-83-014F, Environmental Criteria and Assessment Office, Office of Health and Environmental Assessment, Office of Research and Development, U.S. Environmental Protection Agency, Research Triangle Park, NC 27711.

U.S. Environmental Protection Agency, (1985a), Health Assessment Document for Nickel, EPA/600/8-83/012F, U.S. Environmental Protection Agency, Office of Research and Development, Office of Health and Environmental Assessment, Environmental Criteria and Assessment Office, Research Triangle Park, NC 27711.

U.S. Environmental Protection Agency, (1985b), Measuring Airborne Asbestos Following An Abatement Action, EPA 600/4-85/049, Quality Assurance Division, Environmental Monitoring Systems Laboratory, Office of Research and Development, U.S. Environmental Protection Agency, Research Triangle Park, NC 27711, and Exposure Evaluation Division, Office of Toxic Substances, Office of Pesticides and Toxic Substances, U.S. Environmental Protection Agency, Washington, DC 20460.

U.S. Environmental Protection Agency (1986), Carcinogenicity Assessment of Chlordane and Heptachlor/Heptachlor Epoxide, EPA/600/6- 87/004, Carcinogen Assessment Group, Office of Health and Environmental Assessment, Office of Research and Development, Washington, D.C.

U.S. Environmental Protection Agency, (1987a), "Asbestos-Containing Materials in Schools; Final Rule and Notice," Federal Register, Part III, Environmental Protection Agency, 40 CFR Part 763.

U.S. Environmental Protection Agency, (1987b), Health Assessment Document for Beryllium, EPA/600/8-84/026F, U.S. Environmental Protection Agency, Office of Research and Development, Office of Health and Environmental Assessment, Environmental Criteria and Assessment Office, Research Triangle Park, NC 27711.

U.S. Environmental Protection Agency (1987a), Carcinogenicity Assessment of Aldrin and Dieldrin, EPA/600/6-87/006, Carcinogen Assessment Group, Office of Health and Environmental Assessment, Office of Research and Development, Washington, D.C.

U.S. Environmental Protection Agency, (1988), Radon Measurement Proficiency Program Application and Participation Manual, EPA-520/1-87-022, Office of Radiation Programs, U.S. Environmental Protection Agency, Washington, DC 20460.

Wallace, L. A., (1986), The Total Exposure Assessment Methodology (TEAM) Study: Summary and Analysis, Volume I, Final Report, Environmental Monitoring Systems Division, Office of Acid Deposition, Environmental Monitoring and Quality Assurance, U.S. Environmental Protection Agency, Office of Research and Development, Washington DC 20460.

Wallace, L. A., E. Pellizzari, B. Leaderer, H. Zelon, and L. Sheldon, (1987), "Emissions of Volatile Organic Compounds from Building Materials and Consumer Products," Atmospheric Environment, Vol. 21, No. 2:385-393.

WHO, (1987), "Working Group on Indoor Air Quality: Organic Pollutants Berlin (West) 23-28 August 1987," Draft Final Report, ICP/CEH 026.

Yamate, G., S. C. Agarwal, and R. D. Gibbons, (1984), Methodology for the Measurement of Airborne Asbestos by Electron Microscopy, Environmental Monitoring Systems Laboratory, Office of Research and Development, U.S. Environmental Protection Agency, Research Triangle Park, NC 27711.

Other Noyes Publications

INDUSTRIAL HYGIENE ENGINEERING
Recognition, Measurement, Evaluation and Control
Second Edition

Edited by

John T. Talty
National Institute for Occupational Safety and Health

This book provides an advanced level of study of industrial hygiene engineering situations with emphasis on the *control* of exposure to occupational health hazards. Primary attention is given to industrial ventilation, noise and vibration control, heat stress, and industrial illumination. Other topics covered include industrial water quality, solid waste control, handling and storage of hazardous materials, personal protective equipment, and costs of industrial hygiene control.

The book will serve as a single reference source on the fundamentals of industrial hygiene engineering as related to the design of controls for exposure to health hazards in the workplace. The control of occupational health hazard exposures requires a broad knowledge of a number of subject areas. To provide a text that includes the necessary theoretical foundation as well as the practical application of the theory is a significant undertaking. This text will be a valuable and needed addition to the literature of industrial hygiene, providing the reader with a systematic approach to problem solving in the field of industrial hygiene.

The eight sections of the book are self-contained. Each covers a particular subject area, thus allowing for reference to a single topic area without the need to consult other sections of the book.

CONTENTS

1. INTRODUCTION
Recognition of Health Hazards
Methods for Measuring and Evaluating
  Health Hazards
Human Systems
Industrial Toxicology
Physical Hazards
General Methods of Control Available
Legal Aspects of Occupational Safety
  and Health
2. INDUSTRIAL VENTILATION
Characteristics of Air
Properties of Airborne Contaminants

Principles of Air Movement
Dilution Ventilation
Local Exhaust Ventilation/Make-Up Air
Design of Exhaust Hoods
Principles of Air Cleaning
Air-Cleaning Devices/Air-Moving Devices
Design of Ducts
Principles of System Design
Ventilation Systems Design
Recirculation of Exhaust Air
Correcting for Nonstandard Conditions
Thermal Ventilation Effects
Testing Procedures in the Plant
Environmental Air Pollution
3. THERMAL STRESS
Heat Exchange and Its Effects on Man
Thermal Measurement/Thermal Stress Indices
Methods for Controlling Thermal Exposures
4. SOUND
Physics of Sound
The Ear and the Effects of Sound
Vibration/Noise Control
5. INDUSTRIAL ILLUMINATION
Light
Light and Seeing/Design of a Lighting
  System
Lighting Design
6. NONIONIZING AND IONIZING RADIATION
Principles of Nonionizing Radiation
Control of Nonionizing Radiation
Principles of Ionizing Radiation
Instrumentation
Control of Ionizing Radiation
7. ERGONOMICS
The Worker as the Physical Component
The Worker as the Controlling Component
Design of the Job
Design of the Workplace
8. OTHER TOPICS
Control of Industrial Water Quality
Control of Solid Waste
Purchase, Handling, and Storage of
  Hazardous Materials
Personal Protective Equipment
Costs of Industrial Hygiene Control
Basic Economic Analysis

ISBN 0-8155-1175-2 (1988)　　　　　7"x10"　　　　**831 pages**

HOW TO MANAGE
WORKPLACE DERIVED HAZARDS
AND AVOID LIABILITY

by

Charleston C.K. Wang

This multidisciplinary book was written to help *prevent* the occurrence of workplace derived catastrophe, injury, and illness. The specter of debilitating legal liability for injury and illness from workplace derived causes is ever present for every industry and business. In today's litigious climate, management can ill afford to overlook hazards which may lurk in the workplace. A major workplace derived catastrophe may mean the future viability of a business.

The first part of the book addresses the scientific aspects of the recognition, evaluation, and control of workplace derived hazards. The book next presents a pencilled roadmap to help the reader navigate the complex maze of legal concepts which define and govern workplace derived liability. The book then provides detailed blueprints to enable the creation of the Technical-Legal-Management (TLM) Team for the systematic control of injury, illness, and liability. Today's workplace derived hazards must be correctly *managed* and with excellence. The wisdom of vigilant, efficient physical prevention is emphasized throughout this book. Avoidance of legal liability automatically follows successful physical prevention.

The material in this book, once followed, should result in the effective reduction of workplace derived injury and illness, as well as legal liability. Management, individual employees, and society as a whole, will be the grateful beneficiaries of a successful program to manage and control workplace derived hazards.

A condensed table of contents giving **chapter titles and selected subtitles** is listed below.

ISBN 0-8155-1134-5 (1987)

335 pages

Other Noyes Publications

INDOOR AIR QUALITY CONTROL TECHNIQUES
Radon, Formaldehyde, Combustion Products

by

W.J. Fisk, R.K. Spencer, D.T. Grimsrud
F.J. Offermann, B. Pedersen, R. Sextro

Lawrence Berkeley Laboratory
University of California

Pollution Technology Review No. 144

This book reviews and evaluates existing indoor air quality control techniques. The indoor air pollutants of most concern are radon, formaldehyde, and certain combustion products—nitrogen dioxide, carbon monoxide, carbon dioxide, and various respirable particles. Many techniques exist to control the concentration of these pollutants and other indoor pollutants that are only now being recognized as significant. The purpose of the book is to provide a current review and evaluation of these control techniques.

Environmental concern for air pollution has been largely focussed on questions of outdoor air contamination. Recently, attention has begun to shift to concerns about the quality of air within buildings. In the United States today, people spend only 10-15% of their time outdoors; the rest is spent at home, at work, or traveling in between. Yet existing air quality regulations are based on outdoor conditions, specifically on large-scale, highly visible, outdoor air-pollution sources, such as industrial effluents and vehicle exhaust. Buildings were assumed to shelter occupants from outdoor pollutants, and little thought was given to pollutants generated or trapped indoors. Recent studies have shown that concentrations of certain pollutants indoors exceed standards set for outdoor concentrations. Because health effects are often correlated with exposure over time, it is clear that air quality indoors requires more attention than it has yet received. As an emerging health problem, contamination of indoor air has been linked with a wide variety of building materials and consumer products, as well as strategies that reduce the amount of infiltrating air as a means of promoting energy conservation.

CONTENTS

1. INTRODUCTION
General Context of Indoor Air Quality Problems
Major Indoor Pollutants
Role of Control Techniques in Solution of Indoor Air Quality Problems

2. VENTILATION
Mass Balance Models
Impact of Ventilation Rate on Pollutant Source Strengths

Methods of Ventilation
Concentration-Based Control of Mechanical Ventilation
Ventilation Efficiency
Ventilation of Large Buildings

3. WORKING PRINCIPLES AND BACK-GROUND INFORMATION ON POLLUTANT-SPECIFIC CONTROL TECHNIQUES
Source Control
Removal of Particulate Phase Contaminants
Control of Gas Phase Contaminants
Removal of Radon Progeny by Air Cleaning

4. CONTROL OF RADON AND RADON PROGENY: LITERATURE REVIEW
Source Removal, Exclusion, or Burial
Site Selection
Inhibiting Radon Entry
Radon Progeny Control Through Air Circulation and Particle Control
Summary of Radon/Radon Progeny Specific Control Measures

5. CONTROL OF FORMALDEHYDE CONCENTRATIONS: LITERATURE REVIEW
Formaldehyde Release
Source Exclusion and Removal
Source Modifications
Sealing Wall Cavities
Removal of Formaldehyde from Indoor Air
Summary of Formaldehyde-Specific Control Measures

6. CONTROL OF GASEOUS COMBUSTION PRODUCTS: LITERATURE REVIEW
Nitrogen Dioxide
Carbon Monoxide
Carbon Dioxide
Sulfur Dioxide
Reducing Leakage of Combustion Products to Indoors

7. CONTROL OF RESPIRABLE PARTICLES: LITERATURE REVIEW
Source Exclusion or Reduction
Removal of Respirable Particles from Indoor Air

8. REFERENCES

ISBN 0-8155-1129-9 (1987)

245 pages

acw-458

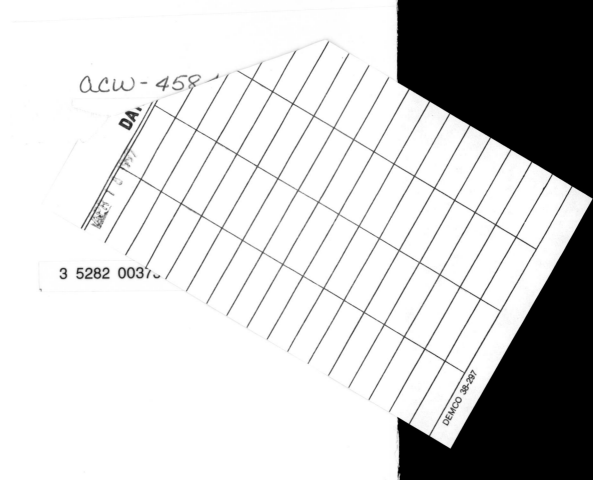

DEMCO 38-297